Lasers in Neurosurgery

Editors:
E. F. Downing (Editor-in-Chief)
P. W. Ascher
L. J. Cerullo
C. R. Neblett
J. H. Robertson
J. M. Tew

Springer-Verlag Wien New York

Edward F. Downing, M.D.
Neurological Institute of Savannah, P.C., Savannah, Ga., U.S.A.

Univ.-Prof. Dr. med. Peter W. Ascher
Neurochirurgische Universitätsklinik, Graz, Austria

Leonard J. Cerullo, M.D.
University Neurosurgery, S.C., Chicago, Ill., U.S.A.

Charles R. Neblett, M.D., P.A.
Houston, Tex., U.S.A.

Associate Professor Jon H. Robertson, M.D.
Department of Neurosurgery, University of Tennessee,
Memphis, Tenn., U.S.A.

Professor John M. Tew, Jr., M.D.
Department of Neurosurgery, University of Cincinnati,
College of Medicine, Cincinnati, Ohio, U.S.A.

With 62 Figures

Library of Congress Cataloging-in-Publication Data. Lasers in neurosurgery/editors, E. F. Downing (editor-in-chief) ... [et al.]. p. cm. Includes bibliographies and index.
ISBN 978-3-211-82067-4 (U.S.) 1. Nervous system—Surgery. 2. Lasers in surgery. I. Downing, E. F. (Edward F.) [DNLM: 1. Lasers—therapeutic use. 2. Neurosurgery—methods. WL 368 L3432]. RD593.L36 1989. 617'.48059—dc19. CIP 88-38982

ISSN 0934-3741
ISBN 978-3-211-82067-4 ISBN 978-3-7091-7607-8 **(eBook)**
DOI 10.1007/978-3-7091-7607-8

Preface

With the exploding progress we are experiencing in the field of lasers in neurosurgery it was felt that a new volume devoted to lasers in neurosurgery is needed. As opposed to other early laser publications which were limited to North American contributors we have decided to publish Lasers in Neurosurgery which presents the findings of neurosurgeons from throughout the world.

The decision to publish all contributions in English, regardless of the native language of the author, makes Lasers in Neurosurgery truly a forum for international neurosurgeons. Our intent is to make available the findings of international neurosurgeons, which are frequently published in less familiar languages, to neurosurgeons beyond the boundaries of the authors' countries.

We hope that neurosurgeons not only in North America and Europe, but throughout the world, will profit by Lasers in Neurosurgery.

November 1988 Edward F. Downing, M.D., F.A.C.S.

Contents

Basic Physics and Biophysics

F. Frank

MBB-Medizintechnik GmbH, Application Research, München,
Federal Republic of Germany

Physics of Lasers

The term *laser* is an acronym for Light Amplification by Stimulated Emission of Radiation. The possibility of laser action was first suggested by Albert Einstein [1] in 1917. In 1954, Charles H. Townes [2] built the forerunner of the laser, a microwave amplifier, to which he gave the name *maser*, an acronym of similar derivation. At about the same time, Basov and Prokhorov [3] independently produced a maser of their own. In 1958, Townes collaborated with Schawlow on a historic paper [4] which laid the theoretical foundation for the laser, then referred to as an optical maser. All of this preliminary work culminated in 1960, when Theodore H. Maiman constructed the first working laser [5] using a rod of crystalline ruby excited by a coaxial helical flashlamp. In rather rapid succession, other lasers were built, notably: the helium-neon, by Javan, Bennett, and Herriott [6] in 1961; the argon-ion by Bridges [7] in 1964; the carbon-dioxide by Patel [8] in 1964; and the neodymium-YAG by Geusic, Marcos, and van Uitert [9] in 1964.

Light amplification by stimulated emission of radiation means we want to achieve light amplification; stimulated emission of radiation is the way to achieve it. The terms light, amplification and radiation are well known. To explain *stimulated emission*, it is necessary to know something about absorption and emission.

Absorption

According to the atomic model of Bohr the electrons in an atomic system stay in separate levels around the protons and neutrons which are assembled in the atomic nucleus. Each of these levels corresponds to a certain energy. If there is an interaction between light-photons with the orbital electrons, the light energy is absorbed and an electron is able to change from one level to the next. This process is called absorption (Fig. 1a).

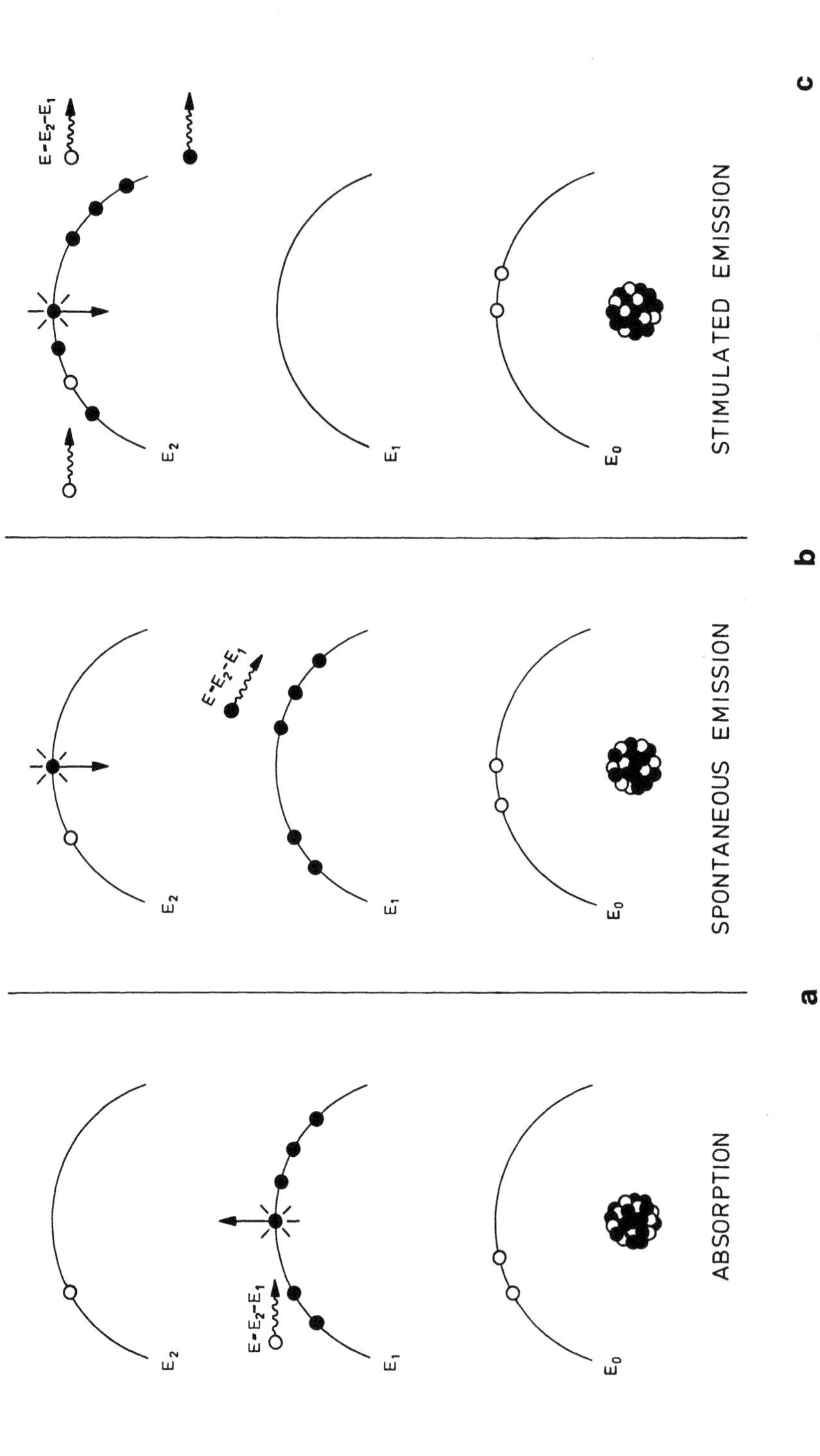

Fig. 1. Diagram of the characteristics of absorption (a), spontaneous emission (b), and stimulated emission of a photon (c) in an atom

Spontaneous Emission

Having absorbed energy in this way, the atom will spontaneously return to a lower energy state after a while. Normally an electron cannot stay in an upper level if the lower level is not completely occupied. The electrons return to the lower energy trajectory. If this decay is radiative, a photon with a wavelength that is proportional to the difference in energy levels is emitted. This process is called spontaneous emission (Fig. 1b).

Stimulated Emission

In the process of stimulated emission, a photon of a specific wavelength hits on an atom in the excited state and causes the electrons to decay to a lower energy state faster than would occur spontaneously. In this case, the incident and the emitted photon travel in the same direction, are in phase, and are of exactly the same wavelength. One photon leads to the emission of a second similar one (Fig. 1c).

Light Amplification

A light amplifier is a lasing medium; that is, a material where stimulated emission is possible, enclosed between two parallel mirrors, the so-called optical resonator. The lasing medium, which may be ions, atoms, or molecules in a solid, liquid, or gas phase, is excited by a pump source that uses light, electrical or chemical energy. In the excited active laser medium photons are being emitted spontaneously in all directions. A good laser medium remains in the excited state for a relatively long period of time. The small population of photons emitted along the axis of the laser resonator will create a cascade of photons. Reflection of photons by the mirrors amplifies the number of photons generated by stimulated emission. One of the mirrors is partially transmitting which permits some of the photons to be emitted from the laser resonator, creating a beam of laser light (Fig. 2).

Properties of Laser Light

Laser light has certain characteristics which distinguish it from ordinary light:

—All emitted photons are of the same wavelengths, therefore the beam will be *monochromatic* (Fig. 3a).

—All the waves, i.e. the photons, are in phase, so the beam will be *temporally coherent* (Fig. 3b).

—All the photons travel in the same direction, so the beam will be spatially coherent or collimated with very *low divergence* (Fig. 3c).

Due to these properties the beam can be focused to achieve high energy densities, making exact work with the laser beam possible.

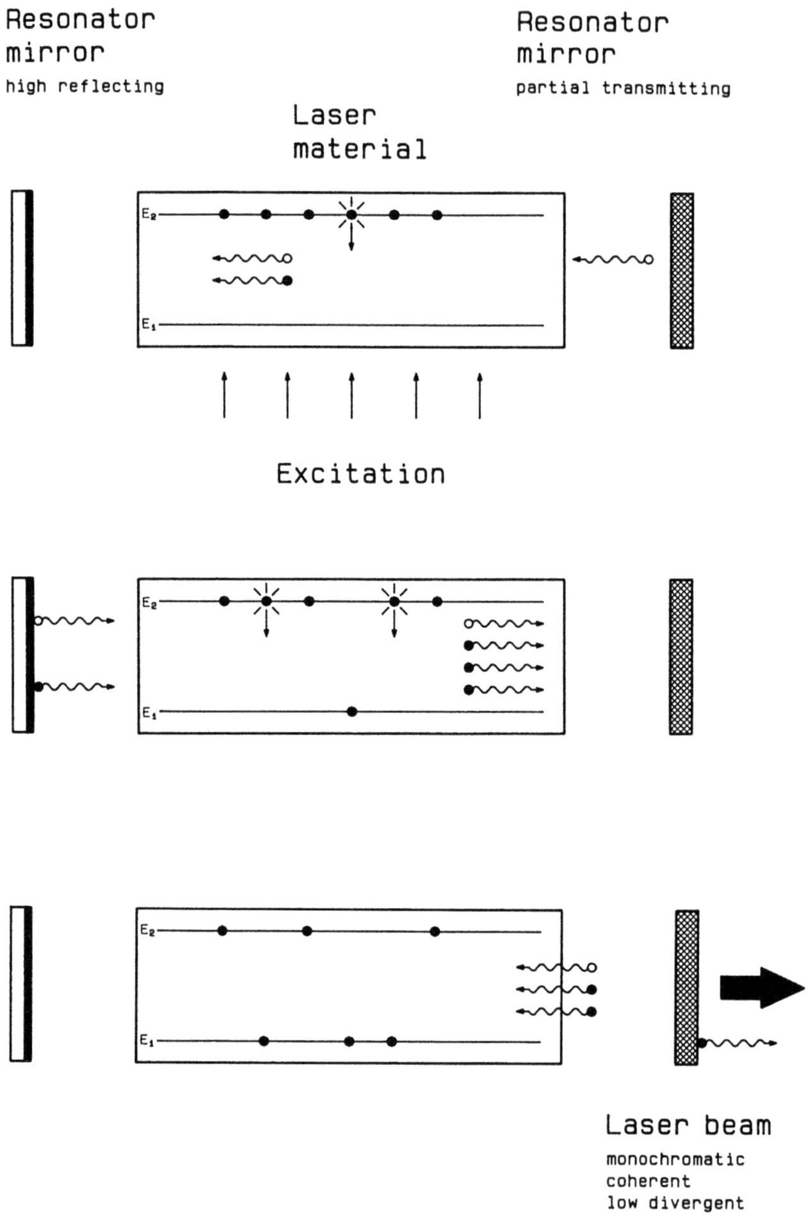

Fig. 2. Principle of laser with the optical resonator, the excitation of the laser material and the light amplification

Laser light can be focused to form very small diameter beams. Focusing the laser beam can increase its intensity by many orders of magnitude. For comparison, an average power density of sunlight is 0.1 W/cm^2, whereas power densities of 100,000 W/cm^2 are easily obtained with surgical laser

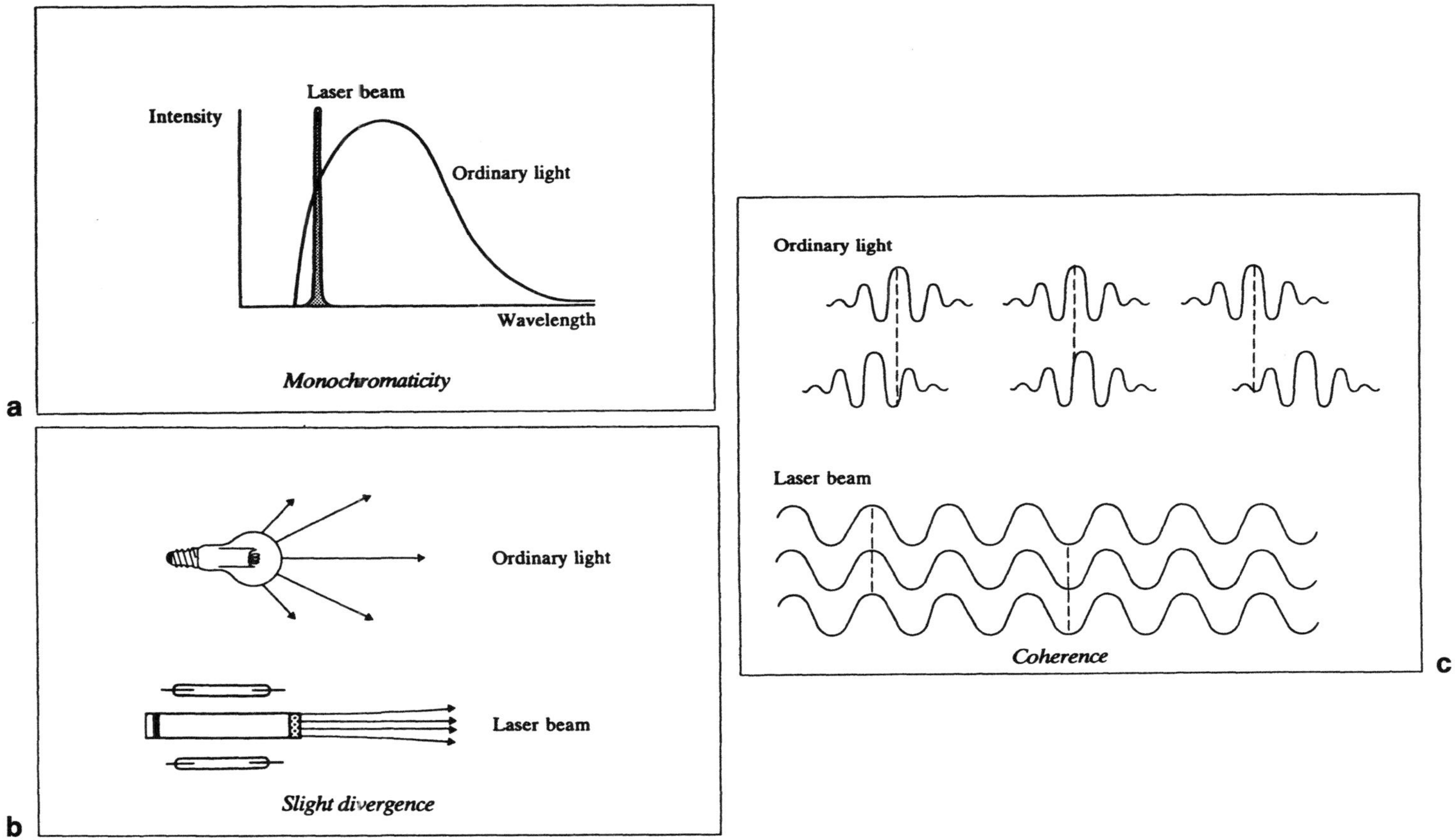

Fig. 3. The spectral band width of laser light (a), the phase relationship of laser light (b), and the divergence of the laser beam (c) compared to an ordinary light source

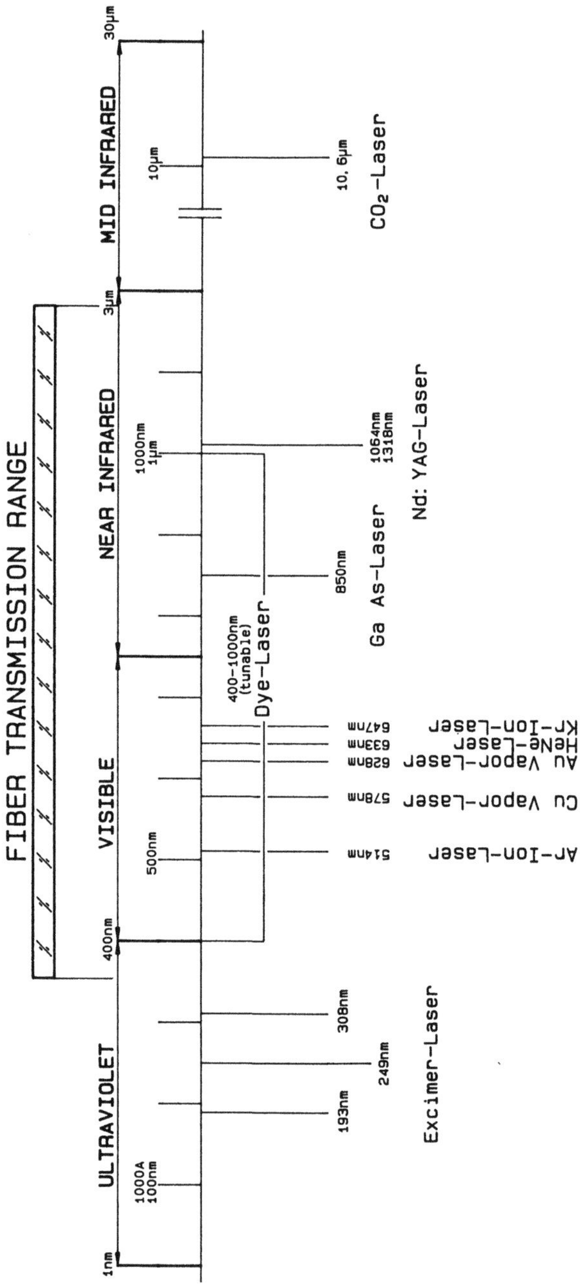

Fig. 4. Different types of lasers and their spectral wavelength

systems. Striking a match produces an energy of 200 J (Ws) of incoherent light. Only with 1 J of coherent light from a ruby laser is it possible, by focusing with a simple lens, to drill a razor-blade.

By using different lasing media, laser systems with different emitted wavelengths can be constructed beginning in the ultraviolet region at about 200 nm and reaching into the infrared up to a wavelength of 10 μm (Fig. 4). In the region of wavelengths between 300 nm and 2,2 μm it is possible to transmit laser light in very thin, but still mechanically robust, quartz glass fibers (0.2–0.6 mm in diameter) on the basis of successive total internal reflection. This offers the endoscopic applications of lasers.

Laser systems differ also with regard to their features concerning duration and power of the emitted laser radiation. In *continuous wave lasers* (cw-mode) with power outputs of up to 10^3 W, the lasing medium is excited continuously. With *pulsed lasers* excitation is effected in a single pulse or in on-line pulses (*free-running mode*). Peak powers of 10^5 W can be developed for a duration of 10 ms (milliseconds) to 100 μs (microseconds). Storing the excitation energy and releasing it suddenly (*q-switch mode* or *mode-locking*) leads to a peak power increase of up to 10^{10}–10^{12} W and a pulse duration of 100 ns (nanoseconds) to 10 ps (picoseconds).

Biophysics of Laser Tissue Interactions

When considering the interaction between laser light and biological tissues, the physical parameters of the biological object must be related to the parameters of the laser light. The degree and extent of the effect depend on the properties of the tissue, which are determined by the structure, water content and blood circulation, i.e. absorption, scattering, reflection, thermal conductivity, heat capacity and density, as well as on the geometry of the laser beam, i.e. its power density, energy content and wavelength (Fig. 5).

Depending on the duration of the laser irradiation on tissue (*interaction time*) on the one hand and on the laser irradiance in surface or volume interaction with tissue (*effective power density*) on the other hand three types of tissue interaction can be distinguished [10, 11]:

—*photochemical* effects (10 s–1,000 s; 10^{-3}–1 W/cm^2)
—*photothermal* effects (1 ms–100 s; 1–10^6 W/cm^2)
—*photoionizing* effects (10 ps–100 ns; 10^8–10^{12} W/cm^2).

Photochemical Effects

With extremely long interaction times and low power densities photochemical transformation occurs by absorption of light with no primary heating of the tissue.

The most important example is the *photosensitized oxidation*. The combined use of laser light and an injected photosensitizer, today mainly he-

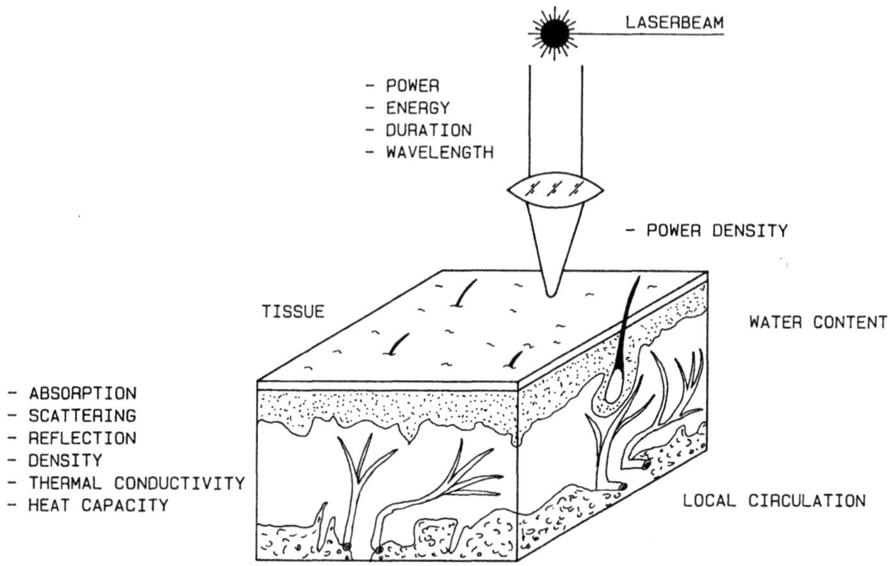

Fig. 5. Diagram of the parameters which determine the effect of laser light in tissue

matoporphyrin derivates (HPD), initiate a cytotoxic process. Most of the tissue is destroyed after excitation of the photosensitizer by laser light. The stimulated sensitizer undergoes a series of intramolecular chemical reactions which lead to the oxidation of various cellular components [12]. The fact that the residence time of HPD in pathologic tissue is longer than in healthy tissue permits selective tumor eradication. In photodynamic therapy use is made of argon pumped *dye lasers* 1 W at 630 nm, cw), and *gold vapour lasers* (10 W at 628 nm, pulsed). HPD has significant side effects. Other photosensitizers for different types of tissue have to be developed.

Biostimulation, mainly for wound healing or pain relief, has to be added to this field too. Systematic studies have not yet given reasonable explanations for the clinically observed improvements [13]. Use is made of *He-Ne laser* (1–5 mW at 633 nm) and *Ga As laser diodes* (5 mW at 850 nm).

Photothermal Effects

With the decreasing interaction time and higher power density the transition to photothermally induced effects begins. The main surgical applications for lasers are based on the conversion of laser light into heat. This thermal effect is broadly applied in surgery for tissue removal and tissue coagulation with sealing of vessels and lymphatics as well as for tissue welding.

The *thermal denaturation* of tissue takes place approximately as follows. To a large extent both the structure and the function of living cells are determined by a wide variety of proteins. These macromolecules have a highly ordered structure which is energetically stable at the body temperature. If the temperature is raised locally to about 50° or more, a certain percentage of the molecules pass into an energetically activated state from which an irreversible transition into the denaturated state takes place. When this happens, the protein molecule loses its spatial arrangement to some extent and with it its power to function in the cell. Depending on the nature of the irradiated tissue, individual thermolabile enzymes may play the leading part in the tissue reaction. There then follows a delayed tissue necrosis, although little or no structural damage to the tissue can be seen immediately after the irradiation.

The degree and the extent of the thermal action depend on the one hand on the optical and thermal properties of the tissue and on the other hand on the laser beam geometry and energy of the incident light.

The most important optical parameter is the wavelength-dependent *absorption* of biological molecules. Since the building blocks of living systems, amino acids, proteins, and nucleic acids, in spite of their great variety are made up of only a few basic elements, some fundamental rules can be formulated for the absorption of optical radiation. The main absorption of biological molecules occurs within the range of wavelength shorter than about 280 nm. The far more molecule-specific vibrational and rotational absorption bands are all in the range of wavelengths longer than 1 μm. Visible laser radiation is hardly absorbed by biological objects. One of the most important exceptions to this rule is the hemoglobin in the red blood corpuscles, and melanin, which is stored as a pigment in the skin and also in large quantities in the pigment epithelium of the retina. A strong absorption in the green spectrum occurs in both substances. The high water content (60%) of most tissue leads to an extensive absorption of infrared radiation. This leads to very efficient energy transfer and heating of the tissue when irradiated with lasers of these wavelengths.

In addition to absorption, *scattering* must be considered as a further optical tissue parameter. Tissue is a highly-structured medium, so that directed optical radiation is completely altered in its spatial distribution due to reflection, refraction, and diffraction. This scattering effect mainly comes into play when absorption is weak.

The thermal properties of tissue, the *heat capacity* and conductivity, can be taken in the first approximation to be the same as those of water. However, estimation of the spread of the energy by *thermal conductivity* is often very difficult when tissue layers of strongly differing structure and complicated geometry are involved, such as the stomach wall, the retina, and the bladder wall, or when blood vessels give rise to a very non-homogeneous removal of energy due to the usually irregular blood flow.

The temperature rise and temperature distribution in tissue exposed to

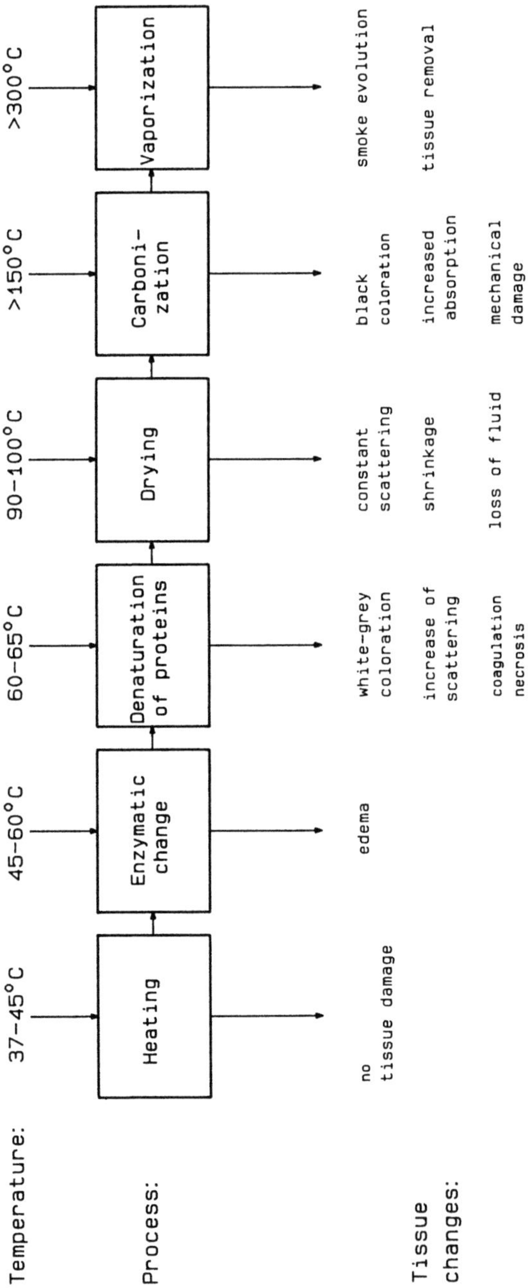

Fig. 6. Thermal tissue alterations following laser irradiation

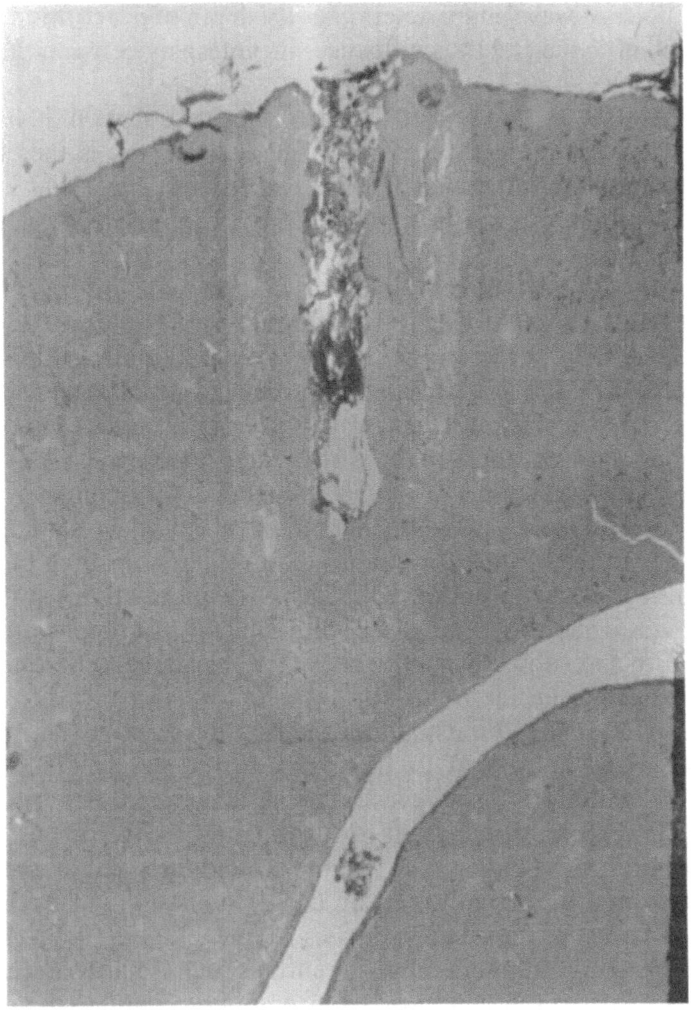

Fig. 7. Cut-like incision in brain tissue with the focused CO_2 laser, density approx. $5\,kW/cm^2$

laser radiation depend on the energy absorbed by the volume of tissue and on the thermal properties of the tissue. The thermal properties determine the temperature increase and the consequent change in temperature distribution following the exposure. According to the temperature in the tissue changes such as discoloration, *coagulation,* shrinkage, *carbonization* and *vaporization* occur. In this process it should be noted that coagulation, unlike vaporization, consumes no additional heat. Vaporization is associated with heat consumption, but despite continued exposure the tissue temperature does not increase during the phase transition (Fig. 6).

The choice of wavelength determines the depth of penetration according to the kind of tissue and thus influences the interplay between the different tissue reactions.

With the *CO_2 laser* (up to 100 W at 10.6 μm) absorption by tissue is the strongest and scattering is negligible. The light energy is therefore completely converted into heat at the tissue surface. For this reason the CO_2 laser as a cutting tool with a small depth of penetration is well suited for removing tissue.

With the CO_2 laser precise incisions can be made in brain tissue, the depth of which is controlled by the duration of exposure. The narrow incision is usually bordered by subarachnoidal bleeding. With sharp focussing (approx. 5 kW/cm^2 at 30 W) a superificially small, very deep, fissure-like lesion with a narrow edematous border is produced (Fig. 7). Short irradiation times of about 0.5 s with a defocussed CO_2 laser (approx. 80 W/cm^2) produce lesions of only a few tenths of a millimeter depth.

The absorption of *argon laser* radiation (up to 10 W, 488nm, 514 nm) is weaker than that of the CO_2 laser radiation. The scattering of the argon laser emission is less pronounced. The penetration is limited due to the selective absorption in hemoglobin and melanin. Its application is thus restricted to indications where removal of the tissue with simultaneous limited coagulation is desired.

The *Nd: YAG laser* emits light in the near infrared range (up to 100 W at 1,064 nm and up to 30 W at 1,318 nm). At the wavelength of 1,064 nm absorption in tissue is very low. Scattering is therefore very pronounced, resulting in uniform distribution of the radiation in the tissue.

Slow heating of a large tissue volume around the point of impingement of the radiation occurs, followed by deep coagulation which slowly progresses. Finally, protoplasm vaporizes at the tissue surface, leading to marked shrinking, although the tissue surface itself is hardly damaged. The shrinking of the tissue combined with uniform coagulation, results in a sealing of blood and lymph vessels. Arteries of up to 2 mm and veins of up to 3 mm in diameter are closed rapidly and reliably. Owing to the shrinkage effect, the coagulation layer causes mechanical compression of the vessels, which results in hemostasis.

With Nd:YAG laser irradiation at 1,064 nm the coagulation volume or damage zone increases according to intensity, then carbonization occurs at the surface, and the tissue is vaporized and removed [14].

Irradiation with a defocused Nd:YAG laser (wavelength 1,064 nm, approx. 180 W/cm^2 at 20 W, exposure time 0.5 s) produces uniform coagulation. Vessels at the brain surface are constricted. The homogeneous volume necrosis is surrounded by a broad edematous boundary. The depth of coagulation depends on the energy applied so that predefined areas can be thermally destroyed to a depth that can be controlled with adequate

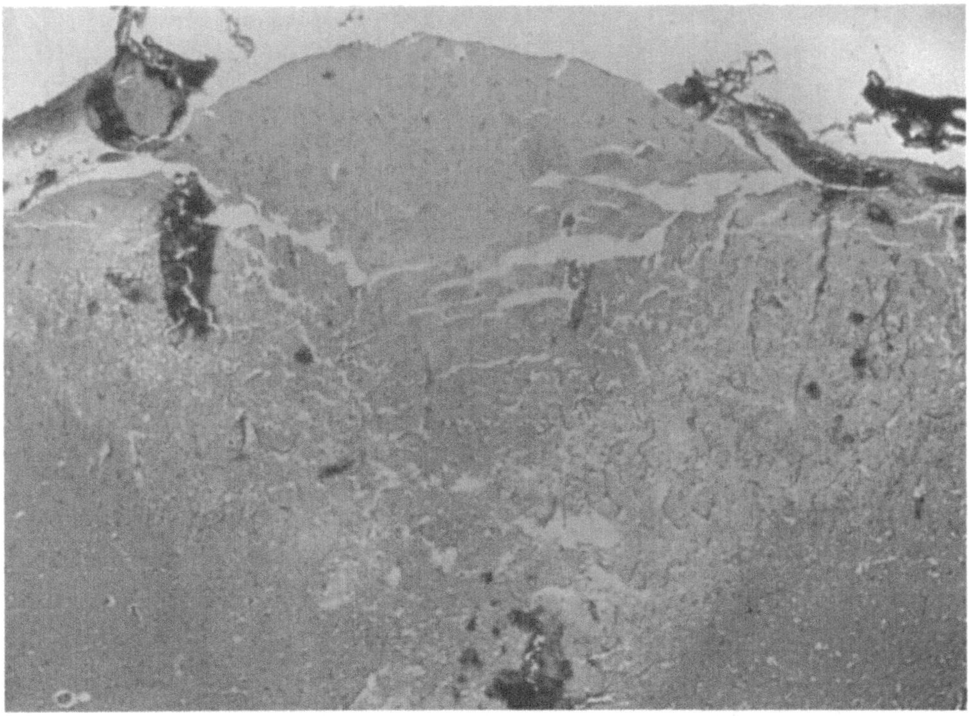

Fig. 8. Wedge-shaped coagulation lesion in brain tissue with the focused Nd:YAG laser, wavelength 1,064 nm, power density approx. 7 kW/cm^2

	ABSORPTION of 0.7% SALINE SOLUTION	EXTINCTION of HUMAN BLOOD
1,06 um	0,07 cm^{-1}	34 4 cm^{-1}
	↓ 10x	↑ 3x
1,32 um	0,73 cm^{-1}	12 3 cm^{-1}

Fig. 9. Absorption of saline solution and extinction of human blood for 1,060 nm and 1,318 nm (Stokes et al.)

precision. With correct dosaging energy densities, the damage can be kept within an area of less than 0.2 mm in depth. Use of a focused Nd:YAG laser beam (wavelength 1,064 nm, approx. 5–7 kW/cm^2 at 30 W) results in rupturing of the cerebral cortex. Owing to scattering of the light the wedge-shaped coagulation zone is limited to 5–6 mm in depth (Fig. 8). Temperature measurements in tissue show that no denaturing occurs in layers deeper than 5–6 mm [15].

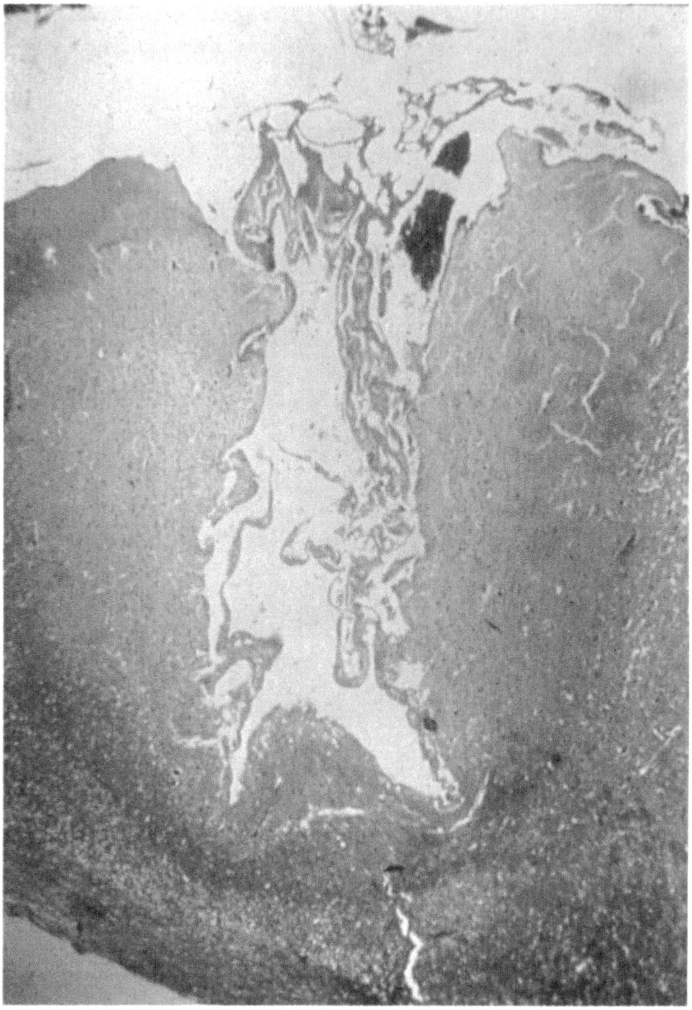

Fig. 10. Loosened necrosis in brain tissue with the Nd:YAG laser wavelength 1,318 nm, power density approx. 1 kW/cm^2

The absorption coefficient of water and saline is approximately ten times higher using the Nd : YAG laser at the wavelength of 1,318 nm than at 1,064 nm. This results in a more efficient conversion of energy into heat in tissue at 1,318 nm. The extinction coefficient (which depends on scattering and absorption) in blood at 1,318 nm is only one third of that at 1,064 nm [16] (Fig. 9). This results in less heat dissipation by blood and deeper penetration in tissue at 1,318 nm.

Particularly in brain tissue a marked loosening of the necrotic tissue is obtained with laser light of the 1,318 nm wavelength resulting in a certain ablative effect using effective power densities of up to 30 kW/cm^2 (Fig. 10). However, it is by no means to be compared with the well-known precise incision obtained with a CO_2 laser. Unlike the CO_2 laser incision, tissue ablation with the Nd:YAG laser at 1,318 nm is distinguished by a clear and effective coagulation zone along the borders of the incision. This provides the possibility of producing narrow, laterally well-demarkated lesions, the depth of which can be controlled by varying the exposure time, an aspect which is important for neurosurgical applications. With a significantly lower effective power density and an interaction time of 0.1 s the resulting sharply defined coagulation can be used for tissue welding [17, 18].

Cortical vessels of the brain, which differ from other blood vessels in some regard (e.g. only thin muscle layer and lack of the elastica externa) could also be coagulated successfully with the Nd:YAG laser by using power densities ranging from 50 W/cm^2 to 600 W/cm^2 at the 1,064 nm or the 1,318 nm wavelength. With cortical vessels of 2 to 3 mm in diameter hemorrhage only occured occasionally after irradiation. The coagulation efficiency of the 1,318 nm laser surpasses that of the 1,064 nm system [19].

The difference in performance leads to different applications and thus indications. The thermal application of lasers in neurosurgery comprises the use of the CO_2 lasers for *cutting* and *vaporizing* tissue, mainly of benign tumors of the midline structures, the base, the forth ventricle and the brain stem as well as the spinal cord [20–22]. Most important is the microsurgical use by adapting the CO_2 laser to the operation microscope. CO_2 lasers in the milliwatt region are used for tissue welding [23].

The Nd:YAG laser is suitable for the *destruction, coagulation* and *shrinking* of tissue specially of vascular lesions like meningeal tumors such as meningiomas of the convexity, particularly in cases with sinus infiltration, as well as basal and intraspinal tumors [24, 25]. This laser is also used in the sellar region [26] and is indicated for AV-malformations and the endoscopic irradiation of intraventricular tumors [28].

New Nd:YAG laser techniques are the endoscopic and interstitial hyperthermia of inoperable cerebral tumors.

The Nd:YAG laser with the wavelength of 1,318 nm especially offers the laser assisted vascular anastomosis (*LAVA*) and the laser assisted nerve anastomosis (*LANA*) as well as the laser assisted aneurysm shrinkage (*LAAS*) [29]. It can also be used for the microsurgical laser assisted tumor extirpation (*LATE*) [30].

A new field is the intravascular application of argon- and Nd:YAG lasers for angiostenotic diseases. To what extent these techniques lead to similarly good results at the carotid bifurcation in the neck as were achieved with thermal laser recanalization of peripheral vessels has still to be investigated.

Photoionizing Effects

When a power density of 10^7 W/cm^2 is exceeded *non-linear effects* result. The high irradiance generates strong electric fields which lead to a *dissociation* or *ionization* of the material involved. Thus laser light is converted into kinetic energy.

The high photon density causes an increased absorption and a direct non-thermal breaking of the intramolecular bonds. This feature is known as photo ablation and has been exploited to produce precise (smaller than 50 μm) non-necrotic cuts by using ArF-, KrF-, XeCl-excimer lasers (10^8 W/cm^2 with 10 ns at 193 nm, 249 nm and 308 nm).

The focusing of even shorter high peak-power laser pulses, creates power densities (10^{10} W/cm^2 for ns pulses and 10^{12} W/cm^2 for ps pulses) which generate such powerful electric fields (10^6–10^7 V/cm), that spontaneous ionization to free electrons and ionized atoms (*plasma*) is induced. When a certain degree of ionization has been reached, a rise in temperature of the plasma follows. The plasma then undergoes a sudden expansion accompanied by a mechanical acoustic shockwave. The shockwave ruptures the tissue structure (*photodisruption*) or disintegrates the targeted material (*photofragmentation*).

Use is made of those ionizing effects in ophthalmology in microsurgical interventions within the eye without any damage of the healthy anatomic structure. Using pulsed q-switched Nd:YAG lasers or pulsed dye lasers and coupling high pulse energies (80 mJ with 10 ns at 1,064 nm or 60 mJ with 1,5 μs at 590 nm respectively) into flexible glass fibers it is possible to smash gallstones in vitro [31, 32] and to fragment kidney and ureter stones in patients [33, 34].

Whether these photoionizing effects achieved with excimer-, pulsed dye- and q-switch lasers can be used in neurosurgery still remains completely open.

References

1. Einstein A (1917) On the quantum theory of radiation. Physikalische Zeitschrift 18: 121
2. Gordon JP, Zeiger HJ, Townes CH (1954) Molecular microwave oscillator and new hyperfine structure in the microwave spectrum of NH_3. Physical Review 95: 282
3. Bassov NG, Prokhorov AM (1954) (article in) Journal of Experimental and Theoretical Physics (U.S.S.R.) 27: 431
4. Schawlow AL, Townes CH (1958) Infrared and optical masers. Physical Review 112: 1940
5. Maiman TH (1960) (report in) Physical Review Letters 4: 564
6. Javan A, Bennett WR, Jr, Herriot DR (1961) Population inversion and continuous optical maser oscillation in a gas discharge containing a Helium-Neon mixture. Physical Review Letters 6: 106

7. Bridges WB (1964) Laser oscillation in singly ionized argon in the visible spectrum. Applied Physics Letters 4: 128
8. Patel CKN (1964) Selective excitation through vibrational energy transfer and optical maser action in N_2—CO_2. Physical Review Letters 13: 617
9. Geusic JE, Marcos HW, Van Uitert LG (1964) Laser oscillations in Nd : doped Yttrium Aluminum, Yttrium Gallium, and Gadolinium Garnets. Applied Physics Letters 4: 182
10. Boulnois J-L (1986) Photophysical processes in recent medical laser developments: a review. Laser in Medical Science 1: 47
11. Müller GJ, Berlien P, Scholz C (1986) Der Laser in der Medizin. Umschau 1986: 233
12. Dougherty TJ (1983) Photoradiation therapy, clinical and drug advances, Porphyrin photosensitisation. Plenum Press
13. Mester E (1980) Laser application in promoting wound healing. In: Koebner HK (ed) Lasers in medicine. Wiley, Chichester, 83pp
14. Frank F, Hofstetter AG, Keiditsch E (1981) Experimental investigation and new instrumentation for Nd : YAG laser treatment in urology. In: Bellina JH (ed) Gynecologic laser surgery. Plenum Press, New York London, 345 pp
15. Edwards MSB, Boggan JE, Fuller TA (1983) The laser in neurological surgery. J Neurosurg: 59: 555
16. Stokes LF, Auth D, Tanaka JL, Gray C, Gulacsik (1981) Biomedical utility of 1.32 μm Nd : YAG laser radiation. IEEE Trans Biomed Eng BME-28: 297
17. Frank F, Beck OJ, Hessel S, Keiditsch E (1986) Comparative investigations of the effects of the Nd : YAG laser at 1.06 μm and 1.32 μm on tissue. Lasers Surg Med 6: 546
18. Schober R, Ulrich F, Sander T, Dürselen H, Hessel S (1986) Laser-induced alteration of collagen substructure allows microsurgical tissue welding. Science 232: 1421
19. Frank F, Beck OJ, Keiditsch E, Schönberger JL, Unsöld E, Wondrazek F (1987) Investigations under neurosurgical aspects with the 1.32 μm Nd : YAG lasers. Laser Med Surg 3: 40
20. Ascher PW (1977) Der CO_2-Laser in der Neurochirurgie. Verlag Fritz Molden, Wien München Zürich Innsbruck
21. Heppner F (1984) CO_2-laser in neurosurgery. Neurosurg Rev 7: 123
22. Ascher PW (1985) Status quo and new horizons of laser therapy in neurosurgery. Lasers Surg Med 5: 499
23. Quigley MR, Bailes JE, Kwaan HC, Cerullo LJ, Brown JT, Lastre C, Monma D (1985) Microvascular anastomosis using the milliwatt CO_2 laser. Lasers Surg Med 5: 357
24. Beck OJ (1984) Use of the Nd-YAG laser in neurosurgery. Neurosurg Rev 7: 151
25. Beck OJ, Frank F (1985) The use of the Nd-YAG laser in neurosurgery. Lasers Surg Med 5: 345
26. Oeckler RTC, Beck OJ, Frank F (1984) Surgery of the sellar region with the Nd-YAG laser. Fortschritte der Medizin 104: 218
27. Tew JM (1987) Treatment of thalamic arteriovenous malformations with the Nd : YAG laser. Laser Brief 9. MBB-Medizintechnik GmbH, München, p 1

28. Auer LM, Ascher PW, Holzer P (1987) Laser-assisted neurosurgical endoscopy. Laser Med Surg 3: 114
29. Ulrich F, Bock WJ, Schober R, Wechsler W (1984) Repair of the carotid artery of the rat with the Nd:YAG laser. Congress of LANSI, Fuschl, 26–30 Sept, 1984
30. Beck OJ, Frank F, Schönberger LJ (1987) Clinical evaluation of the 1.32 μm Nd:YAG laser. Laser Med Surg 3: 110
31. Ell Ch, Wondrazek F, Frank F, Hochberger J, Lux G, Demling L (1986) Laser-induced shockwave lithotripsy of gallstones. Endoscopy 18: 95
32. Simon W, Hering P (1987) Laserinduzierte Stoßwellenlithotripsie an Nieren- und Gallensteinen (in vitro). Laser und Optoelektronik 1: 33
33. Hofstetter A, Frank F, Keiditsch E, Wondrazek F (1985) Intracorporale, laserinduzierte Stoßwellen-Lithotripsie. Laser Med Surg 1: 155
34. Hofstetter A, Schmeller N, Pensel J, Arnholdt H, Frank F, Wondrazek F (1986) Harnstein-Lithotripsie mit laserinduzierten Stoßwellen. Fortschr Med 35: 32

Address for correspondence: Dr. Frank Frank, MBB-Medizintechnik GmbH, Application Research, P.O.B. 80 11 68, D-8000 München 80, Federal Republic of Germany.

The Treatment of Arteriovenous Malformations of the Brain with the Neodymium: YAG Laser

J. M. Tew Jr, W. D. Tobler, and **M. Zuccarello**

Department of Neurosurgery, University of Cincinnati, College of Medicine, Cincinnati, Ohio, U.S.A.

Introduction

The introduction of the Neodymium: yttrium-aluminum-garnet (Nd: YAG) laser is the most recent technological adjunct advocated for the treatment of arteriovenous malformations (AVMs). Beck reported that Nd: YAG laser is effective for coagulation of brain tissue [1, 2]. Subsequent preliminary reports concerning the clinical effectiveness of the Nd: YAG laser for coagulation and induction of hemostasis associated with vascular tumors [9, 29], aneurysms [8], and AVMs [7, 8, 31, 34] have varied from enthusiastic to disappointing in content. Blood selectively absorbs the energy emitted by this laser wavelength and thrombosis of vessels can be effectively induced without imparting substantive damage to adjacent brain tissue [32]. Because the operative removal of AVMs of the brain and particularly those arising from the central core remain one of neurosurgery's major unresolved technical challenges, we have directed our efforts toward technical developments of the Nd: YAG laser as a mode of therapy. We agree with Wharen and colleagues that refinements are necessary [34]. However, our experience in a substantial number of cases of complex AVMs confirms the safety of the device and indicates that it can be of inestimable benefit in the surgical extirpation of some malformations.

The following report presents our findings and conclusions resulting from these preliminary experiences.

Surgical Technique

During the course of this evaluation, two Nd: YAG lasers were used. Each emits continuous wavelengths of energy at incident powers up to 100 watts. The laser light is delivered by 50 mm focused handpieces, nonfocused fibers closely applied to the malformed vessels and, more recently and precisely,

Table 1. *Cerebral arteriovenous malformations treated with Nd: YAG laser*

Case	Age (yrs) &sex	History	Neurological signs	Location	Surgery outcome Size	ICH	IVH	Excision	Complication	Neurological function
CS/1	28/M	SAH (× 2)	Aphasia, right hemiparesis	Left temporo-parietal (para-ventricular)	S	+	+	I stage	None	Excellent
JW/2	71/M	SAH	Left homonymous hemianopsia, left hemiparesis	Right parieto-occipital (superficial)	L	+	+	I stage	None	Good
CR/3	34/M	Seizures	Dysmetria; left homonymous hemianopsia	Right parieto-occipital (superficial)	S	–	–	I stage	None	Good
JD/4	27/F	Headache, seizures	None	Left Sylvian fissure	M	–	–	II stages	None	Excellent
HJ/5	19/F	SAH (× 2)	Left homonymous hemianopsia	Right parieto-occipital (medial)	L	–	–	I stage	Postop hematoma	Excellent
JJ/6	15/M	SAH (× 4)	Right homonymous hemianopsia	Left temporo-parieto-occipital (para-ventricular)	L	+	+	II stages	None	Excellent

CW/7	48/M	SAH (×2)	Dysphasia, right hemiparesis	Right thalamus (postero-superior)	S	+	+	I stage	None	Fair
GL/8	20/M	SAH	Dysphasia, right hemiparesis, right homonymous hemianopsia	Left temporo-parieto-occipital (paraventricular)	L	+	+	I stage	None	Excellent
MS/9	29/M	SAH (×4)	Left dystonia, left homonymous hemianopsia	Right thalamus (postero-superior)	M	+	+	II stages	None	Good
SC/10	15/M	Seizures	None	Right frontal	M	–	–	I stage	None	Excellent
LT/11	32/F	Seizures	Normal	Right frontal	L	–	–	I stage	Postop hematoma	Good
PM/12	37/F	SAH (×2)	Mild ataxia	Right cerebellum	S	–	–	I stage	None	Excellent
SH/13	21/F	SAH	Right cortical sensory deficit	Right parietal	M	+	–	I stage	None	Good

Key: M = Male; F = Female; SAH = Subarachnoid hemorrhage; ICH = Intracerebral hemorrhage; IVH = Intraventricular hemorrhage; S = Small (<2.5 cm); M = Medium (2.5–5 cm); L = large (>5 cm)

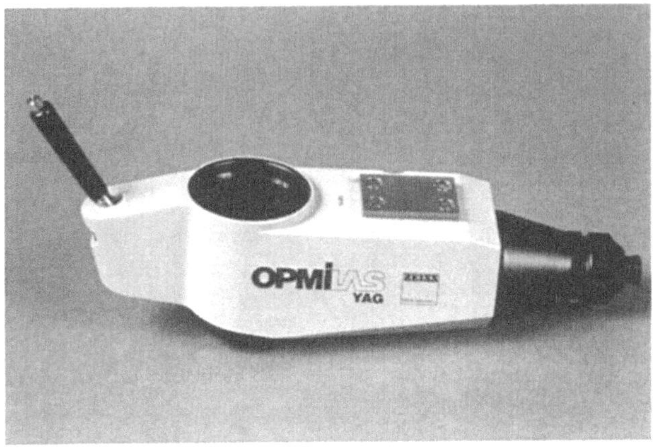

Fig. 1. Nd:YAG micromanipulator attaches to base of microscope for precise microsurgical application (Carl Zeiss, Inc. Thornwood, N.Y.)

by a focusing micromanipulator attached to the surgical microscope (Fig. 1). A coincident helium-neon laser is used as an aiming device. Intermittent pulses of variable length up to 9.9 seconds at incident powers of 5 to 75 watts are used to obliterate vessels of the AVM. The laser beam is used at varying size focal points depending on the diameter of blood vessels. All resections were peformed under the operative microscope. General anesthesia with barbiturate protection and hypotension induced by nitroprusside is used regularly during the operative procedures. Complete resection of all AVM was confirmed by postoperative angiography. The neurologic and psychometric status of all patients was confirmed by repeated examinations.

Case Summaries

The Nd: YAG laser was first used at our institution to vaporize a small residual malformation on the left lateral ventricular wall in February 1984. Our total series of arteriovenous malformations operated on with the Nd: YAG laser now numbers more than 50 cases. Table 1 shows some selected reports of cases, distribution of the lesions, and surgical outcome. There were eight males and five females with an average age of 30.5 years. There were nine right-sided and four left-sided lesions. Three of the lesions were located primarily in the thalamus. Outcome is classified as excellent, good, and fair. Excellent outcome was achieved in seven cases; the patients are neurologically normal, or demonstrate improving neurological function relative to the preoperative status. Good outcome was achieved in five cases. The neurological deficits of these patients were unchanged after surgery. There was one fair result in which the patient had an increased hemiparesis and hemisensory deficit postoperatively. The follow-up period ranges from one to 24 months.

There were two postoperative complications, both intracranial hematomas. The first (case number five) occurred 18 hours after surgery and was due to inadequately controlled blood pressure. The second (case number eleven) occurred in the recovery room, and an inadequately controlled basal dural vessel was responsible. Both patients were immediately re-explored and recovered completely.

The following cases have been chosen to illustrate the techniques and features of our experience in the treatment of arteriovenous malformation with the Nd : YAG laser.

Case Number Four—J.D.

This 27-year-old woman had chronic headaches and temporal lobe seizures. A large temporal-Sylvian malformation fed by branches of the middle cerebral artery was identified in the dominant hemisphere (Fig. 2). She was referred for treatment because of concern that neurological disability would complicate surgical treatment.

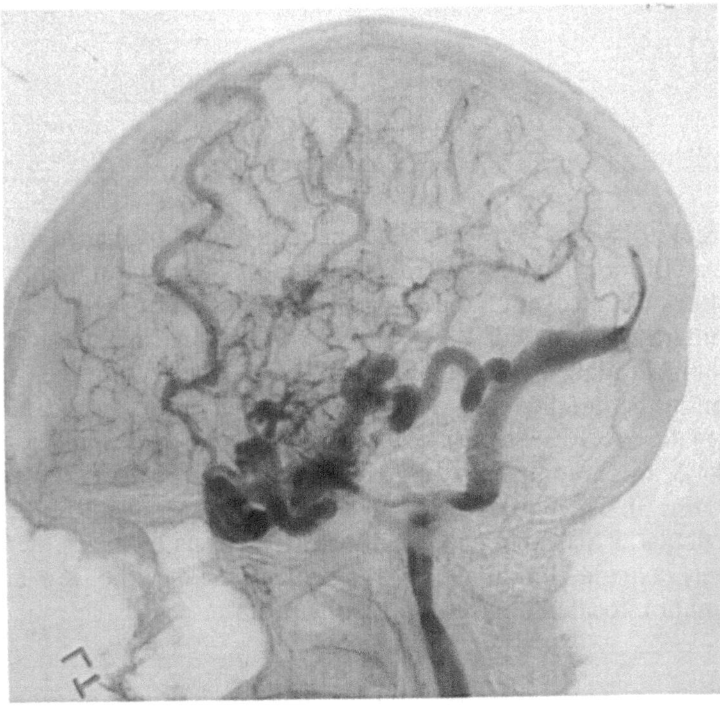

Fig. 2a. Large dominant temporal-Sylvian AVM fed by two large middle cerebral artery branches. Lateral view

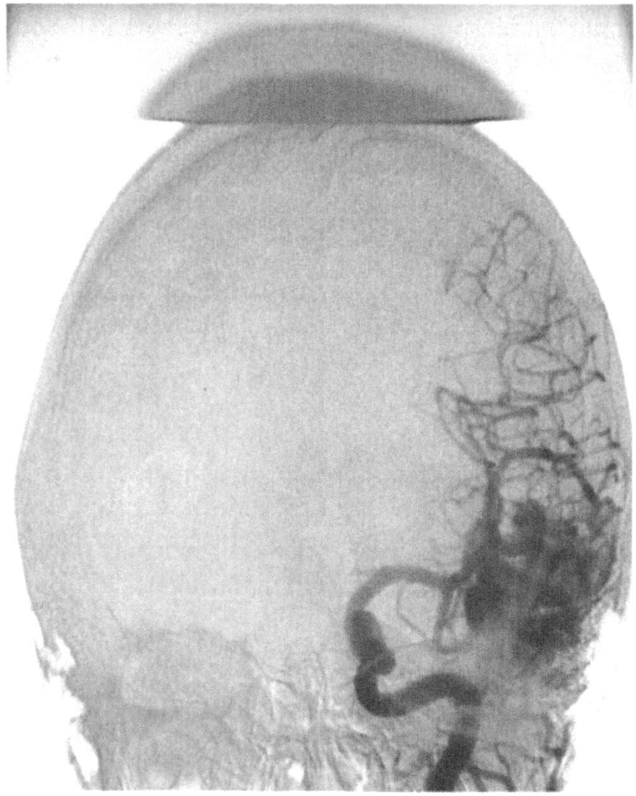

Fig. 2b. AP view

A craniotomy was performed and the patient was awakened for cortical stimulation to localize the speech area. A prominent dural component of the malformation covering the floor of the temporal fossa was coagulated and obliterated with Nd:YAG energy. This maneuver created intense hemicranial headache which could be relieved by injection of the trigeminal ganglion in the middle fossa. The nidus of the malformations was treated with the Nd : YAG laser to reduce its bulk. Complete dissection and excision of the AVM was accomplished with a combination of bipolar and Nd : YAG energy (Fig. 3). The patient made an excellent recovery with no detectable speech deficit. This experience demonstrated the capacity of the laser to effectively coagulate dural vessels and radically reduce the flow in a large malformation located adjacent to eloquent brain.

Case Number Six—J.J.

A 15-year-old male presented eight years previously with a subarachnoid hemorrhage and was found to have a large left temporal-occipital AVM

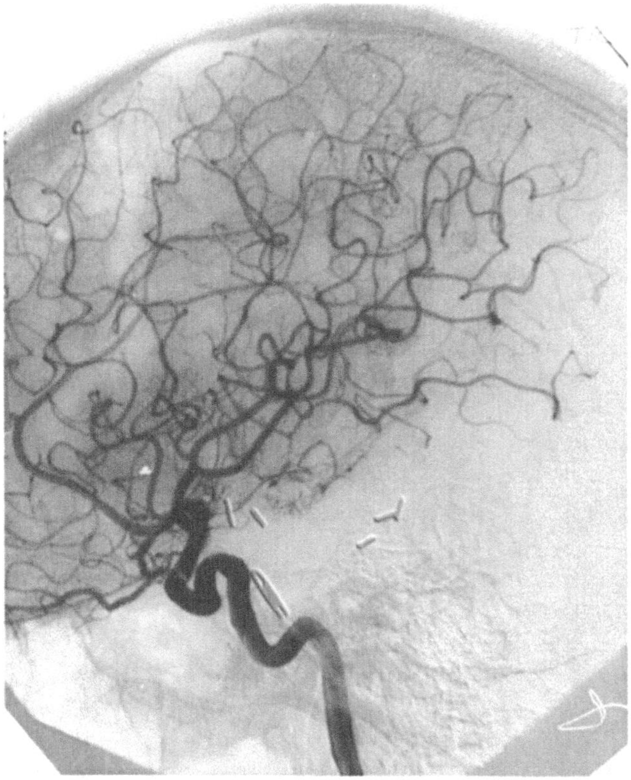

Fig. 3a. Postoperative angiogram shows complete excision of the lesion. Lateral view. Metallic clips are on the dura and a single large draining vein

fed by branches of the middle and posterior cerebral arteries (Fig. 4). He underwent proton beam irradiation in 1978. In 1983 and twice in early 1985 the child experienced major recurrent subarachnoid and intracerebral hemorrages. Neurological examination demonstrated a complete right homonymous hemianopsia. Preoperative embolization with polyvinyl alcohol particles was undertaken and the flow in the posterior cerebral artery was reduced. At surgery with Nd : YAG laser energy the malformation was irradiated and coagulated, and subsequently excised. Residual malformation in the thalamus required a second stage procedure to coagulate deep choroidal feeders and excise the densely fibrotic malformation from the pulvinar (Fig. 5). Postoperative angiogram showed complete removal of the AVM (Fig. 6). The patient's only neurological deficit was residual homonymous hemianopsia. We learned that a fibrotic, toughened malformation could be coagulated and excised by adjusting the power and focus of the Nd : YAG laser. Consequently, a difficult malformation could

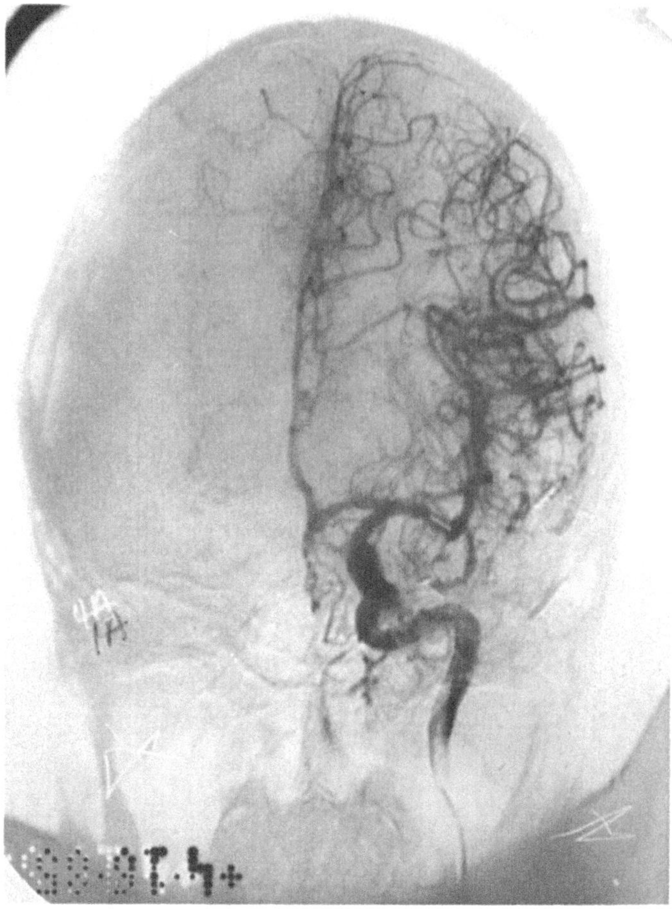

Fig. 3b. AP view

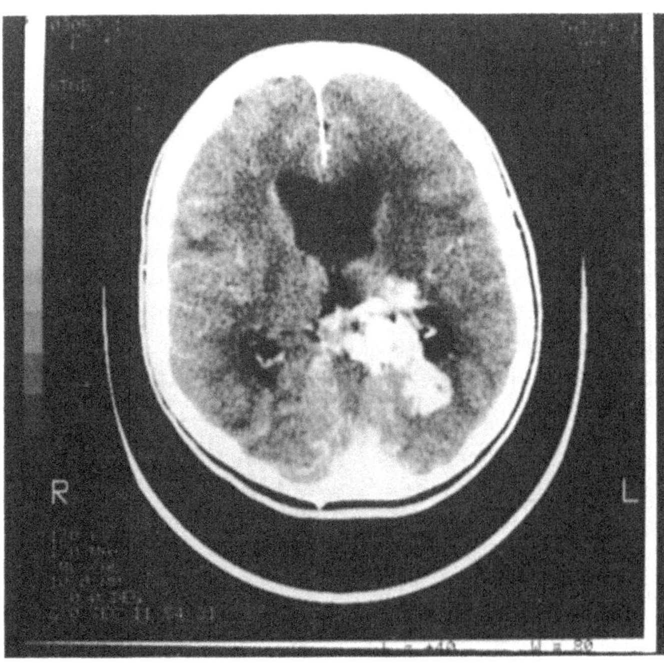

Fig. 4a. Contrast CT shows deep thalamic extension of the AVM

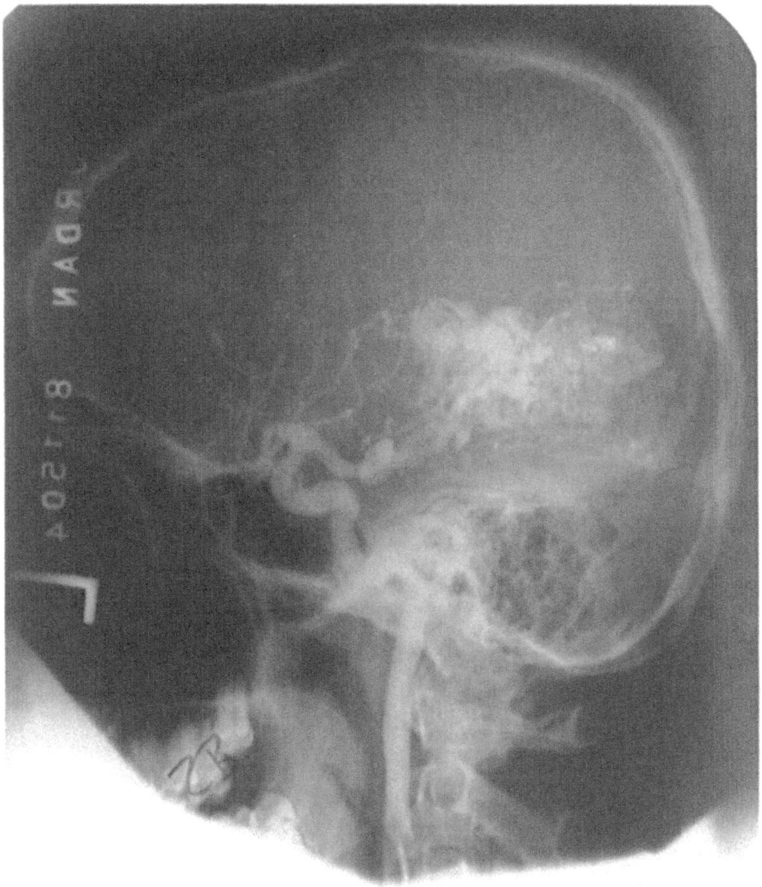

Fig. 4b. Lateral angiogram shows deep malformation fed by branches of middle and posterior cerebral arteries

be completely excised from the cerebrum, choroid plexus and thalamus. We were very encouraged by the results of this experience and are convinced that the Nd: YAG laser can be helpful in removal of complex malformations of the thalamus and central brain core.

Case Number Seven—C.W.

A 48-year-old male with two prior intraventricular and thalamic hemorrhages presented with a right thalamic AVM fed by anterior choroidal, middle and posterior cerebral arteries (Fig. 7). Examination after the most recent hemorrhage disclosed mild hemiparesis and severe thalamic sensory deficit. The lesion was exposed via the transcallosal approach and the AVM

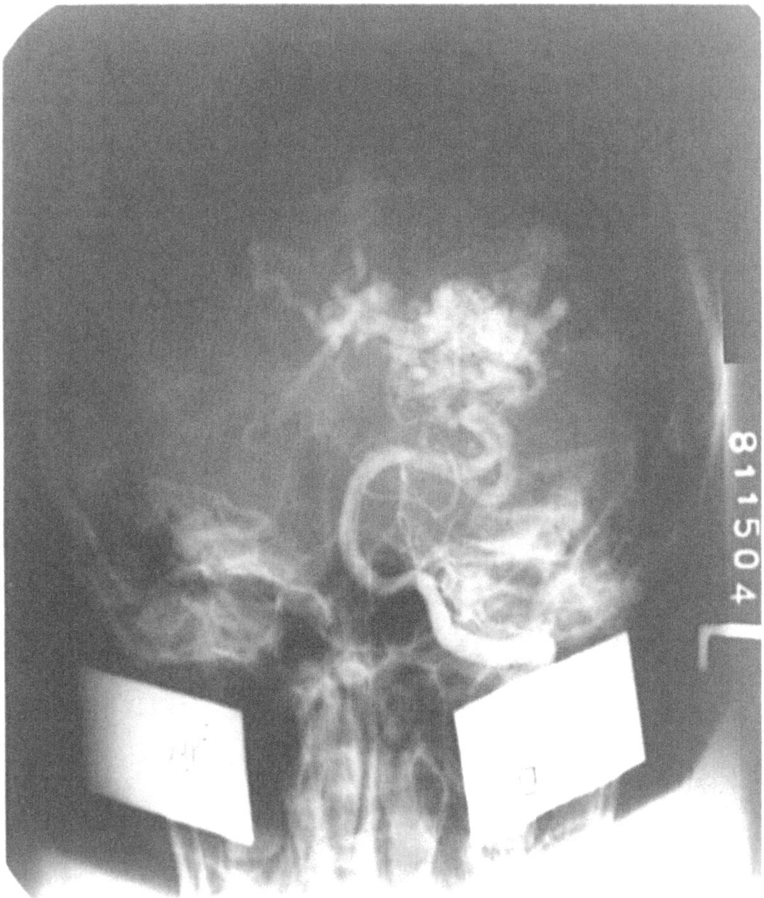

Fig. 4c. AP view showing posterior cerebral contribution to the AVM

was identified in the middle of the thalamus. The AVM was resected using a combination of bipolar electrocautery and Nd:YAG energy. A clip was used to occlude the large posterior cerebral feeding vessels. The postoperative angiogram showed no residual lesion (Fig. 8). The patient has a persistent postoperative hemiparesis and only mild worsening of his sensory deficit. We were very encouraged by this experience, although the patient suffered a significant worsening of the motor deficit.

Case Number Nine—M.S.

This 29-year-old male was found to have a right thalamic AVM in 1974, at the time of an initial intrathalamic hemorrhage. Subsequently the patient experienced serious hemorrhages in 1979, 1984, and 1985. He had left-

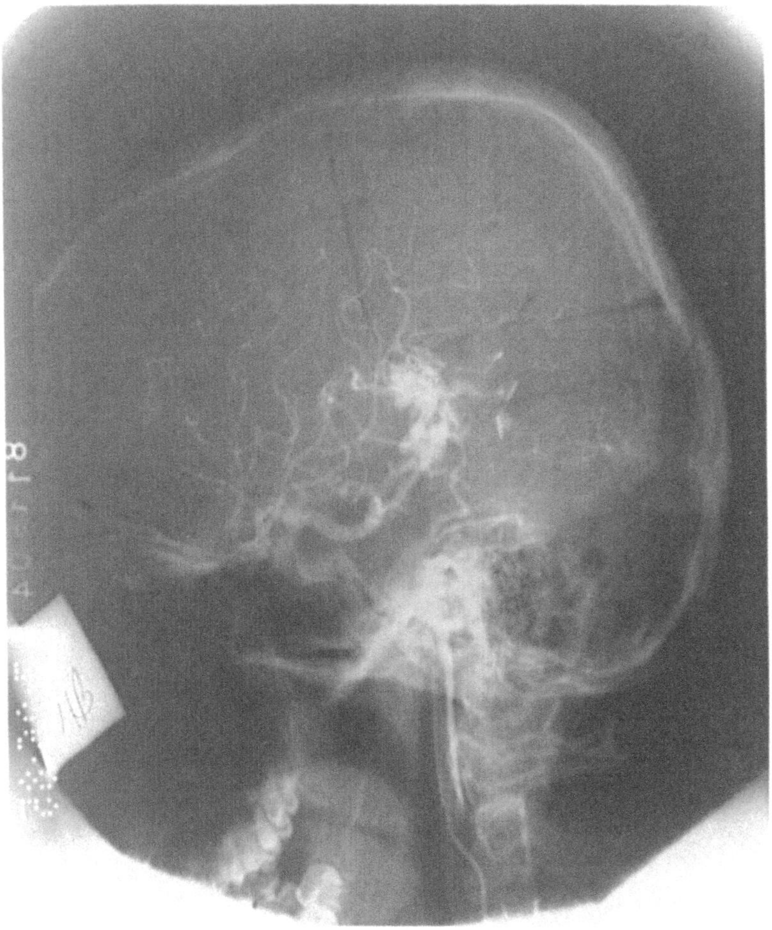

Fig. 5a. Residual malformation, in the deep pulvinar region, after first procedure. Lateral view

sided dystonia, dysmetria, hemiparesis, and spasticity. The angiogram showed extensive primary feeding vessels, principally from anterior choroidal and striato-perforate arteries (Fig. 9). Note the remarkable size and number of the vessels entering the lesion from deep surface. There was minimal supply from the posterior thalamo-perforate vessels.

A decision for surgical extirpation was elected because of increasingly severe, repetitive hemorrhages. The lesion was exposed through a transventricular approach, and was vaporized with the Nd:YAG laser. The feeding vessels were occluded and coagulated with bursts of 20 watts defocused energy. Postoperatively, the patient had increased dystonia, and a partial hemianopsia. However, in the follow-up period, the patient improved to his preoperative condition, and returned to full employment as

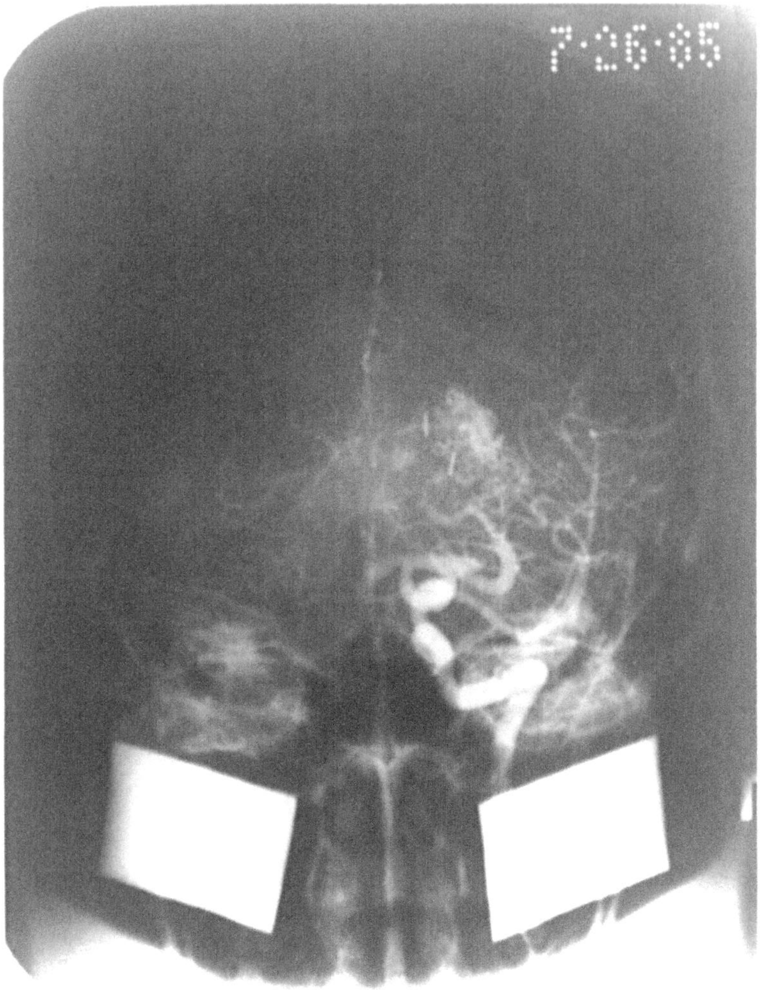

Fig. 5b. AP view of residual malformation

a computer programmer. Angiogram shows no residual malformation (Fig. 10).

This case provides definitive experience that a complex thalamic malformation can be effectively eliminated by this laser-assisted technique. The ability to occlude large, deep-seated vessels on the inferior surface of the thalamus was demonstrated. A transventricular approach was chosen because of the large size of the malformation and the presumed need to gain wider access to the lesion. In the future, transcallosal exposure will be adequate unless a large portion of the malformation lies in the atrium of the lateral ventricle.

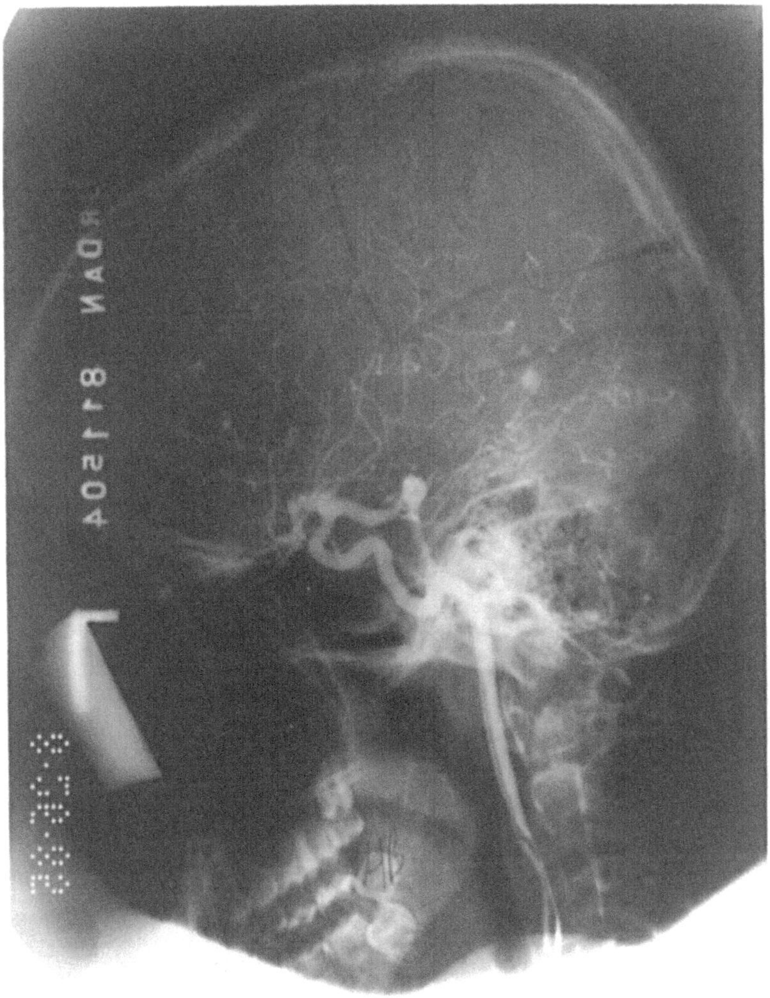

Fig. 6a. Second postoperative angiogram shows no residual malformation. Lateral carotid injection

Discussion—History of AVM Surgery

Arteriovenous malformations of the brain were first described by Luschka in 1854, Virchow in 1867, and Steinhill in 1895. Giordano in 1905 and Krause in 1908 reported their operative experience with these lesions. In 1928 Dandy reported eight cases, only one of which was cured by ligature of a simple arteriovenous fistula [4]. It was not until 1941 that the first successful cases of total excision were presented by Penfield and Erikson. A monograph by Olivecrona and Riives dealt with the subject in detail and reported on 43 cases of complete excision with four deaths. In 1969,

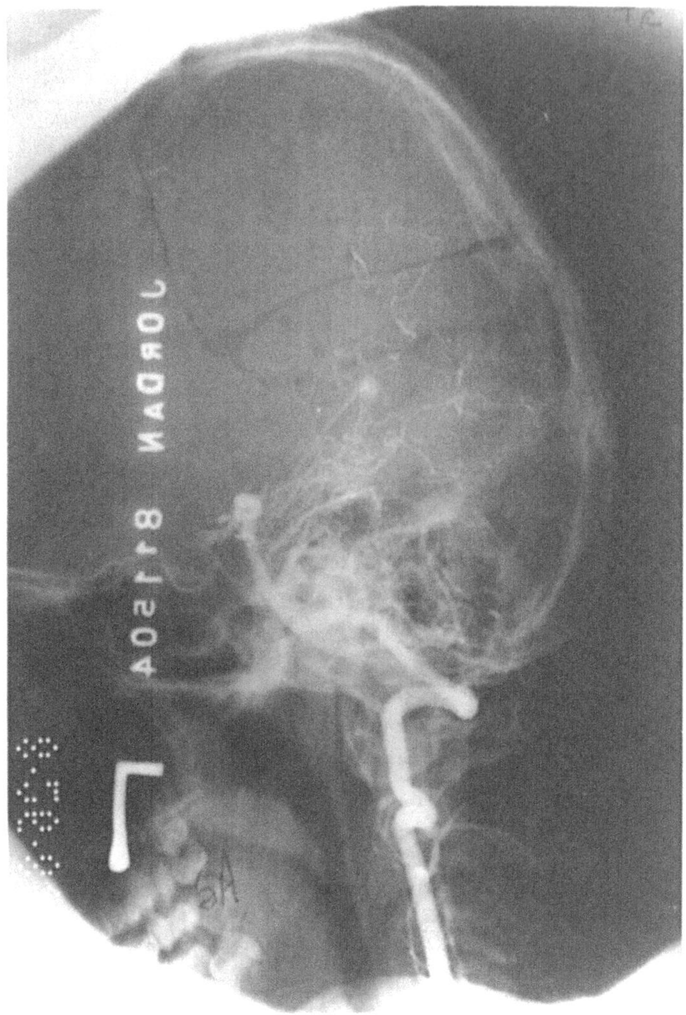

Fig. 6b. Lateral view. Vertebral injection

Yasargil reported excellent results in 11 of 14 cases, 13 of which had been previously considered inoperable. In this landmark paper, he reported the first use of the operating microscope for arteriovenous malformations. Microsurgical technique has proven to be a major technological advancement which today is routinely employed by most neurosurgeons. In the subsequent decade and a half, several reports indicate the safety and efficacy of surgical treatment of arteriovenous malformations [3, 6, 23, 25, 26, 27, 30]. The morbidity and mortality rates associated with microsurgical excision are superior to the natural history of this disease [21]. These cumulative experiences indicate a compelling case for surgical treatment of

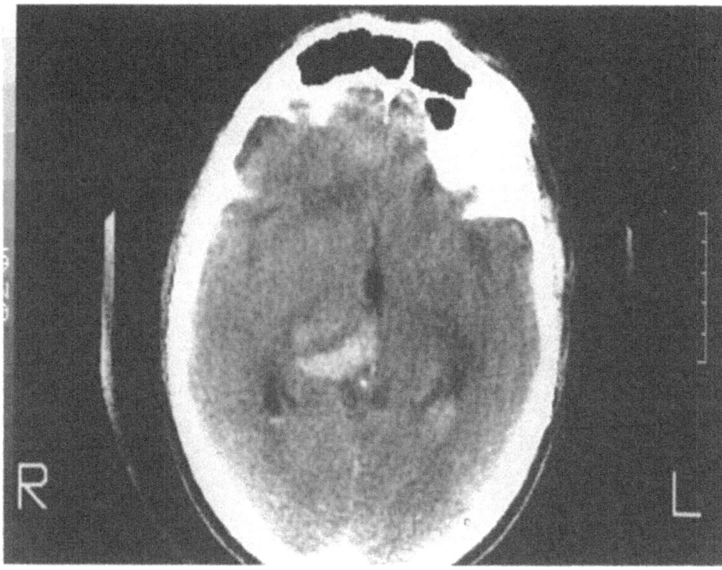

Fig. 7a. Noncontrast CT documenting hemorrhage from right thalamic AVM

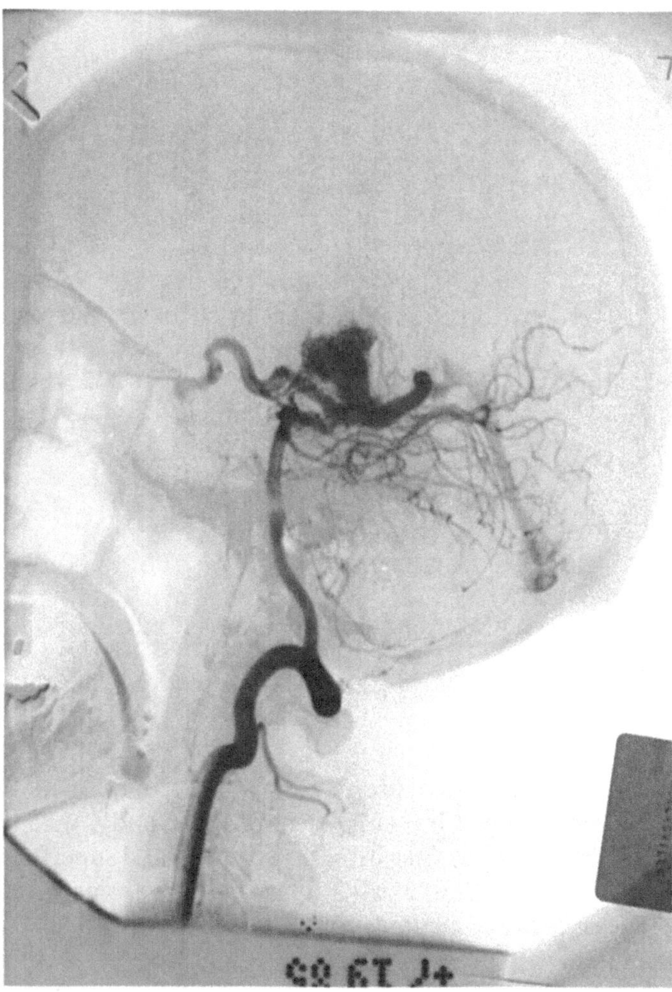

Fig. 7b. Thalamic AVM. Posterior choroidal feeders on lateral vertebral injection

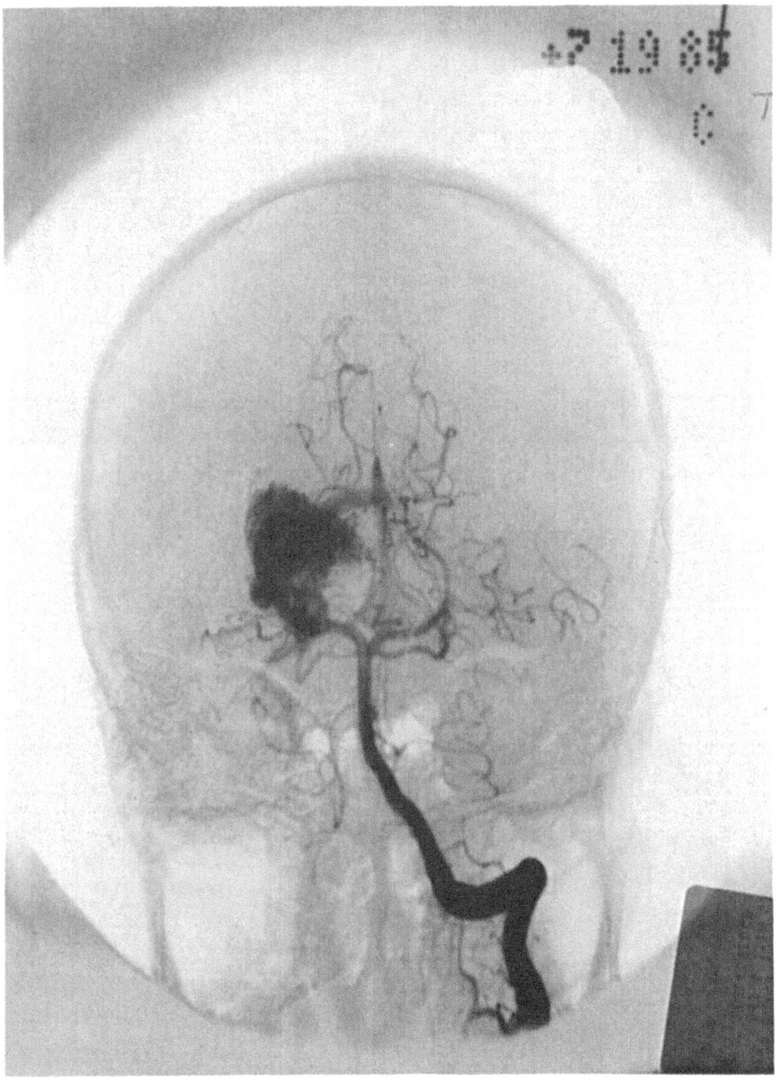

Fig. 7c. Thalamic AVM. Choroidal feeders on AP vertebral injection

most lesions which involve the cerebral hemisphere and present with seizures or headache.

To be effective, total surgical excision must be accomplished. This concept was first acknowledged by both Cushing and Dandy, who described the surgical alternatives. Surgical extirpation of small and superficial lesions can be easily accomplished. The risks for patients with larger and more deeply-seated lesions are sometimes considered too great to advocate surgical treatment [10].

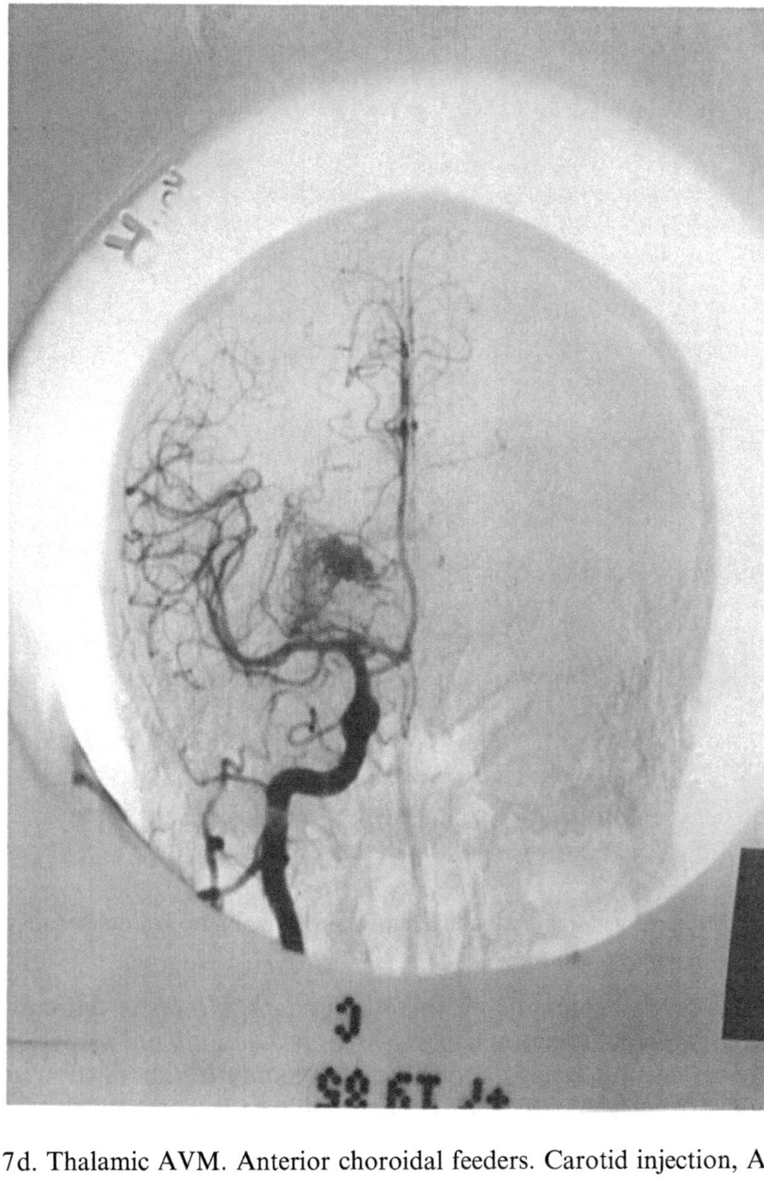

Fig. 7d. Thalamic AVM. Anterior choroidal feeders. Carotid injection, AP view

Particulate occlusion of the malformation by flow-directed embolization of the feeding vessels was a radical new technique introduced by Luessenhop in 1960 as an alternate method of treatment for difficult malformations [19, 20]. From this seminal experience the field of vascular interventional neuroradiology has developed. Today, detachable balloons, rapidly polymerizing adhesive, and other thrombogenic particles are used to obliterate malformations [5, 11, 24, 30, 35]. Total elimination of complex malfor-

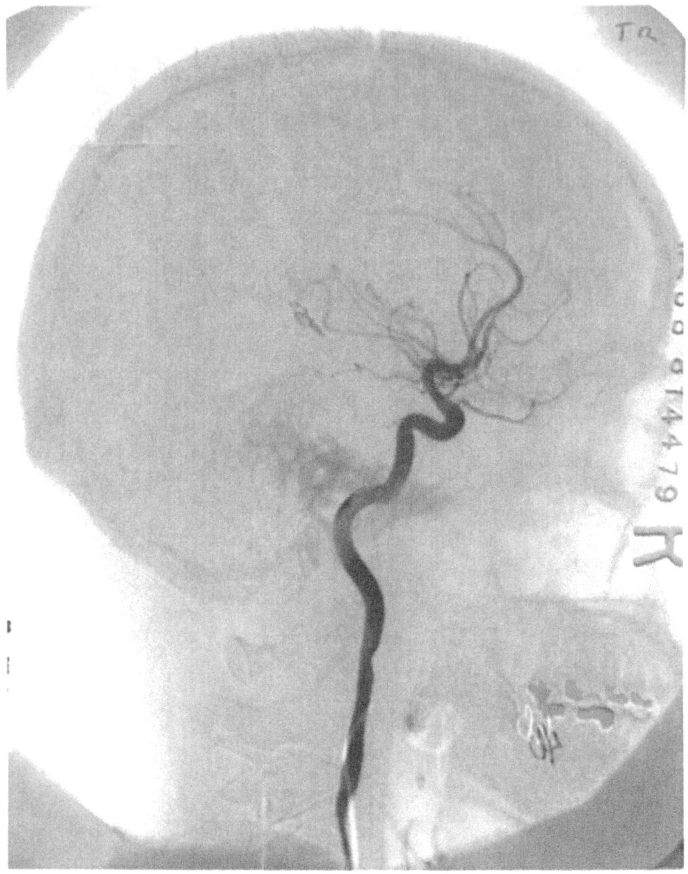

Fig. 8a. Postoperative lateral carotid angiogram showing no residual malformation

mations is rarely achieved with these techniques. However, difficult cases may be converted into more reasonable surgical lesions and palliation may be achieved in some circumstances. We do not know if embolization reduces the incidence of hemorrhage or epilepsy [30]. At our institution, the role of vascular interventional techniques is to treat lesions of the dural vessels and to reduce the flow and size of surgically accessible lesions immediately prior to surgical extirpation. It is anticipated that the role of vascular interventional neuroradiology techniques will expand beyond the adjunctive role as technical expertise and experience is gained.

Another nonsurgical method of treatment of inaccessible arteriovenous malformations which is potentially so important as to merit surgical attention is radiation therapy. In 1969, Steiner proposed the use of Leksell's unit in which gamma rays from 179 Cobalt rod sources could be collimated and focused stereotactically to give necrotizing radiation in a spherical 50%

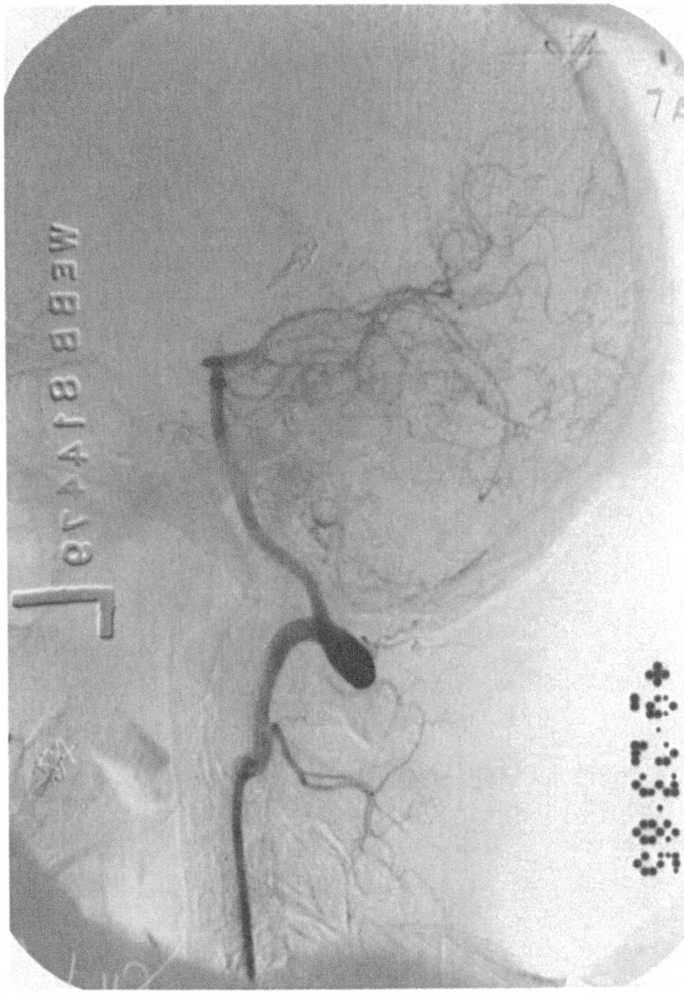

Fig. 8b. Lateral vertebral angiogram

isodose distribution of 2.5 cm anywhere in the brain. He has reported favorable results in the control of arteriovenous malformations [28].

Kjellberg reports his experience using the Bragg peak proton beam from the Harvard accelerator stereotactically focused on these lesions [17]. Both Steiner and Leksell, and Kjellberg, are encouraged, and both groups still advocate strongly surgical excision where it is felt to be reasonably safe. Radiotherapy of AVMs is reserved for small, unresective lesions of the brainstem and basal ganglia, and the long-term results are not yet fully reported.

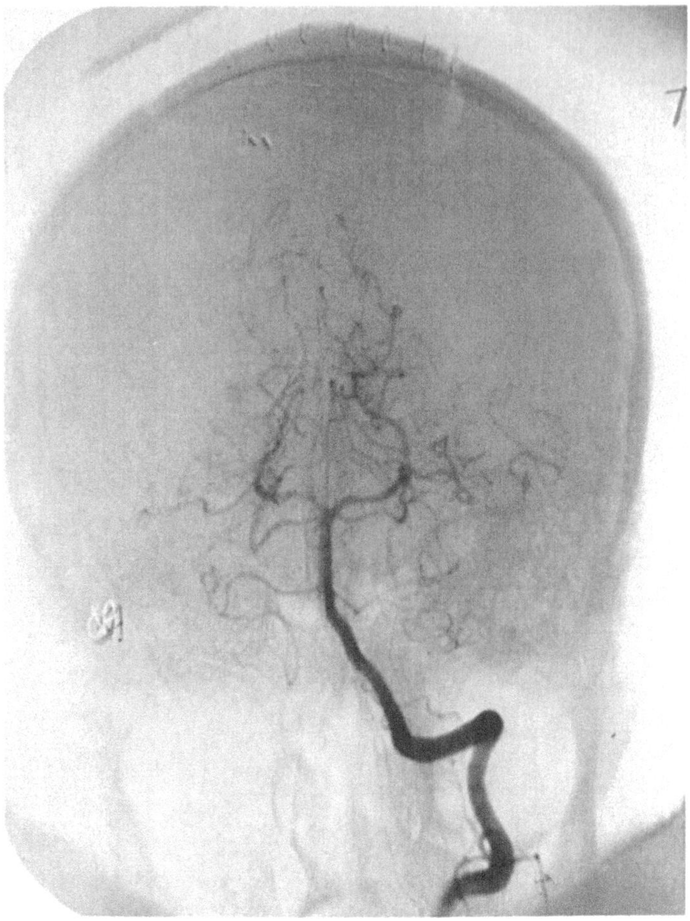

Fig. 8c. AP view

The Role of Lasers in Neurosurgery

After the first laser was developed by Maiman in 1960, all sectors of the business, industrial, and scientific world were eager to explore application for this revolutionary technology. Twenty-six years later, lasers continue to change the world and have clearly had an impact on the practice of neurosurgery.

The first students of lasers created lesions and defined the histological changes brought about by ruby and then carbon dioxide lasers. Stellar first used the laser on a patient with a glioblastoma in 1969. In 1976, Heppner and Ascher, and Takizawa, in Japan, began using the carbon dioxide laser for tumors of the central nervous system. Since then, refinements in laser technology, and improving technical experience have led to the increasing

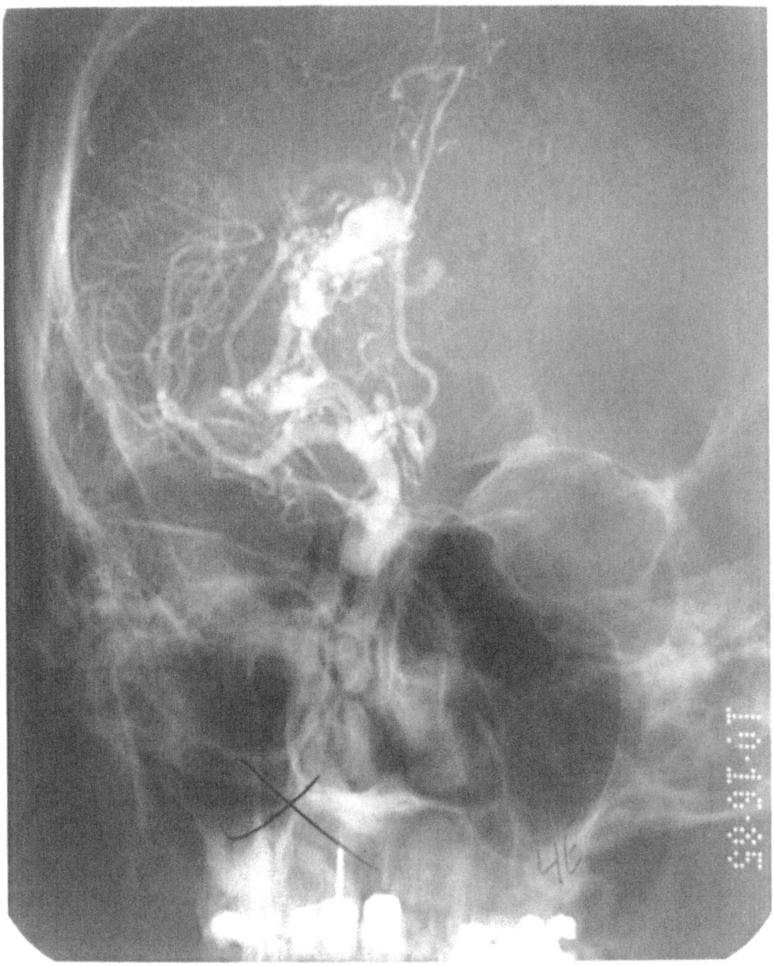

Fig. 9a. Thalamic AVM. Carotid injection, oblique view

acceptance and application of carbon dioxide lasers for tumor surgery. The carbon dioxide laser is well suited to tumor extirpation because of its high power output, surface absorption irrespective of pigmentation, vaporization and hemostatic properties [13, 15, 31]. This powerful laser may be focused to a point for use as a scalpel. Observation of the hemostatic and coagulative properties encouraged us to apply the CO_2 laser in the treatment of small AVMs. However, we found that the CO_2 laser was largely ineffective in the treatment of larger vascular lesions because the superficial absorption characteristics and the cutting effect of this laser often cause disruption and bleeding of a vascular lesion. Therefore, we began to explore the possible application of Nd : YAG energy for the extirpation of vascular malformations.

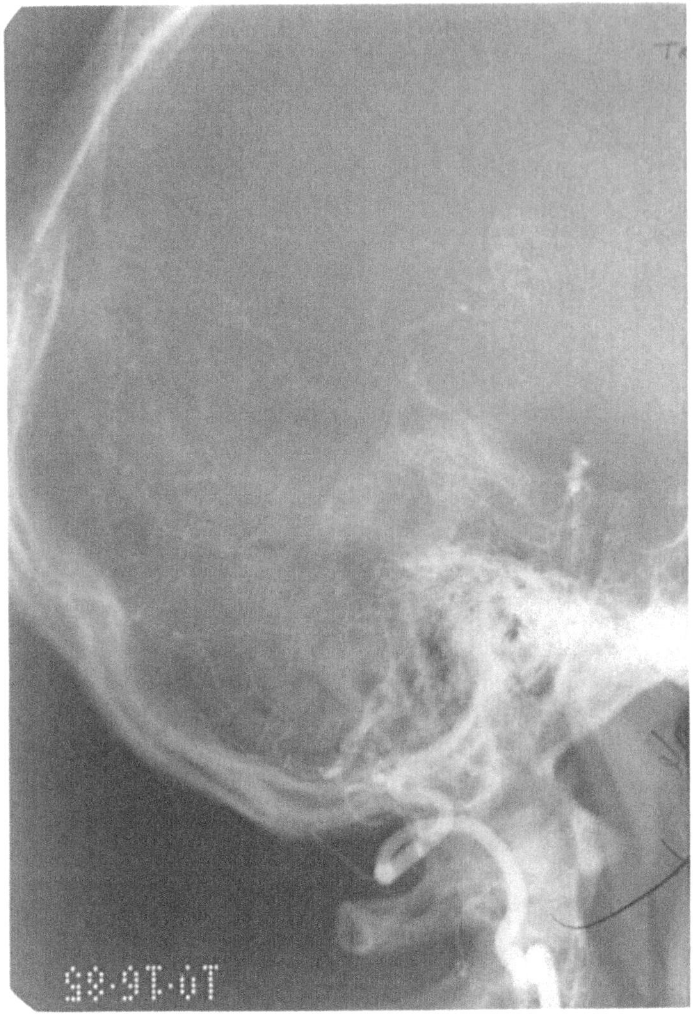

Fig. 9b. Thalamic AVM. Vertebral contribution. Lateral view

The Nd: YAG Laser in Vascular Neurosurgery

The Nd: YAG laser possesses wavelength characteristics that make it better suited for treatment of vascular lesions. Located in the infrared wavelength of 1.06 microns, the energy is scattered deeply below the surface of the tissue to which it is applied. This results in a deeper heating of the tissue than occurs with a CO_2 laser (Fig. 11). Furthermore, the wavelength is preferentially absorbed by red pigmented tissue, a factor causing selective effects on vascular structures [12, 18, 31, 33].

In 1976, Beck began experiments with the Nd: YAG laser, and reported

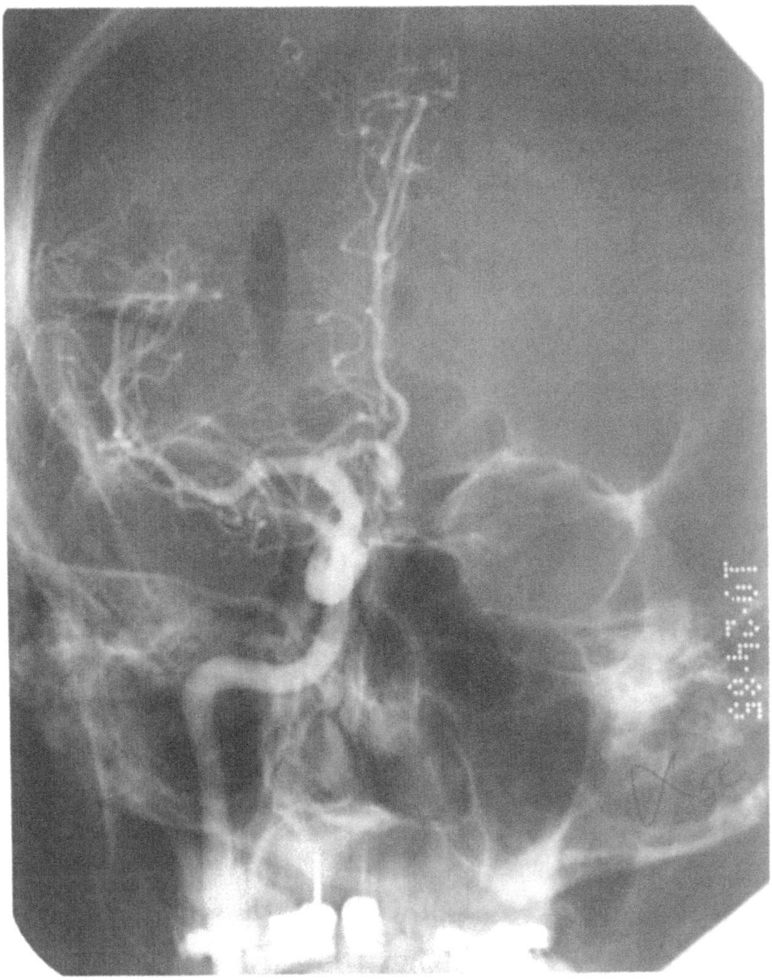

Fig. 10a. Postoperative angiogram documenting no residual malformation. Carotid injection. AP view

that the Nd:YAG was effective for coagulation of vascular tumors [2]. Beck first reported using the Nd:YAG laser as an adjunct in AVM surgery in 1980 [1]. This was followed by Fasano's report of six arteriovenous malformations and five aneurysms treated with either argon or Nd:YAG energy [8]. There is no detailed discussion of the technique of laser excision of the malformation, but the author indicates the Nd:YAG laser is most effective in coagulating thin-walled veins. More recently, Wharen and Sundt reported their experience in the resection of ten parenchymal arteriovenous malformations [34]. Five of the cases were described in some detail. Nine of the ten cases were one-stage procedures. Angiographic demonstration

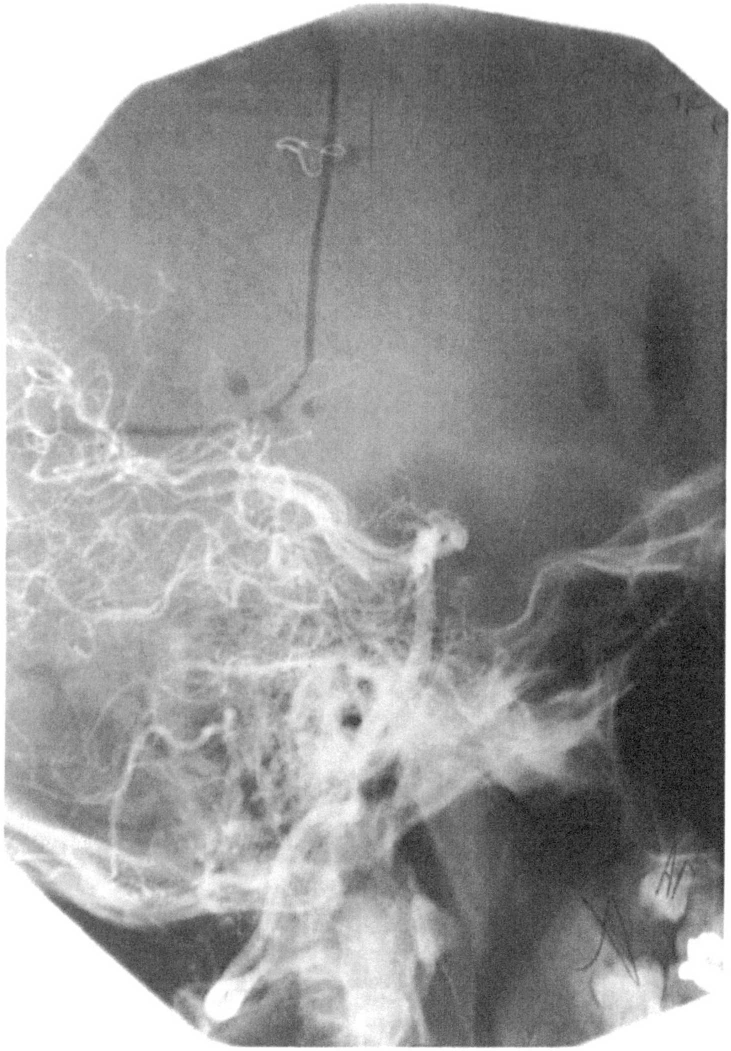

Fig. 10b. Postoperative angiogram. Vertebral injection. Lateral view

of complete excision was provided for all cases. They concluded that the Nd: YAG laser was helpful for 1) defining the plane between the AVM and the brain; 2) coagulating any dural component of the AVM, and 3) achieving hemostasis in the bed following resection of the lesion. They did not feel the laser was beneficial in achieving control of the small thin-walled vessels on the deep surface of the malformation. They further commented that vessels with higher flow are more difficult to coagulate. In spite of these stated drawbacks, they maintained that the Nd: YAG laser

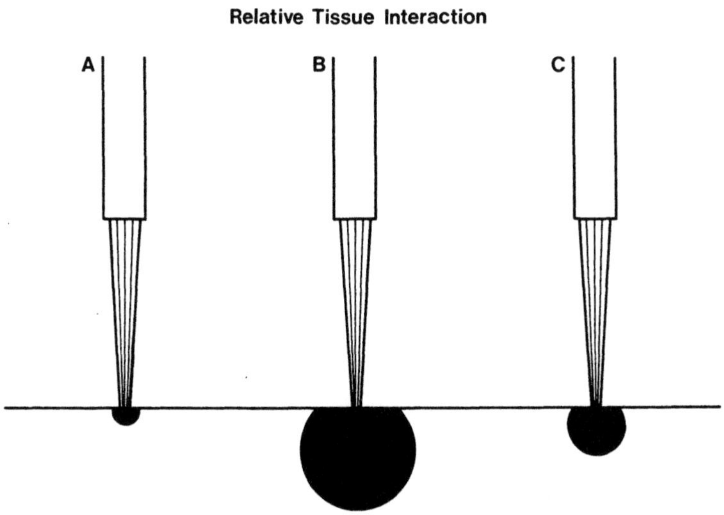

Fig. 11. Relative scatter effect of laser energy. The carbon dioxide laser has minimal scatter (*A*). The Nd:YAG laser has maximum tissue scatter (*B*). The argon laser is intermediate (*C*)

could safely be applied in AVM surgery and that the equivocal results could be improved by further technical advances.

Our series of cases represents the largest reported series of arteriovenous malformations treated with the Nd:YAG laser as the primary mode of resection. We agree that the Nd:YAG laser is helpful for dissecting the plane between nidus and normal brain and for achieving hemostasis in the bed of resection. More importantly, direct radiation of the AVM nidus causes shrinkage of the AVM and reduces the blood flow through it. This effect occurs as a result of the scatter and heating effect which causes contraction of the collagen fibers and results in narrowing of the vessel lumen. This is demonstrated in the histological section of AVM treated with Nd:YAG energy (Fig. 12). The vessel wall has contracted after radiation, resulting in occlusion. A small amount of thrombus is present. As pointed out by Wharen and Sundt, high flow through a vessel reduces the heating effect on the vessel lumen, and results in failure to achieve shrinkage or thrombosis of the vessel and it may not be possible to coagulate the end of a bleeding vessel. Therefore, it is essential to stop the flow, at least temporarily. These factors are responsible for the inability of Nd:YAG radiation to shrink and totally occlude most AVMs with a "no touch" technique as can be achieved with carbon dioxide laser vaporization of brain tumors.

Our personal experience with AVMs confirms all of Wharen et al. opinions. Although collaboration with the laser industry has been effective,

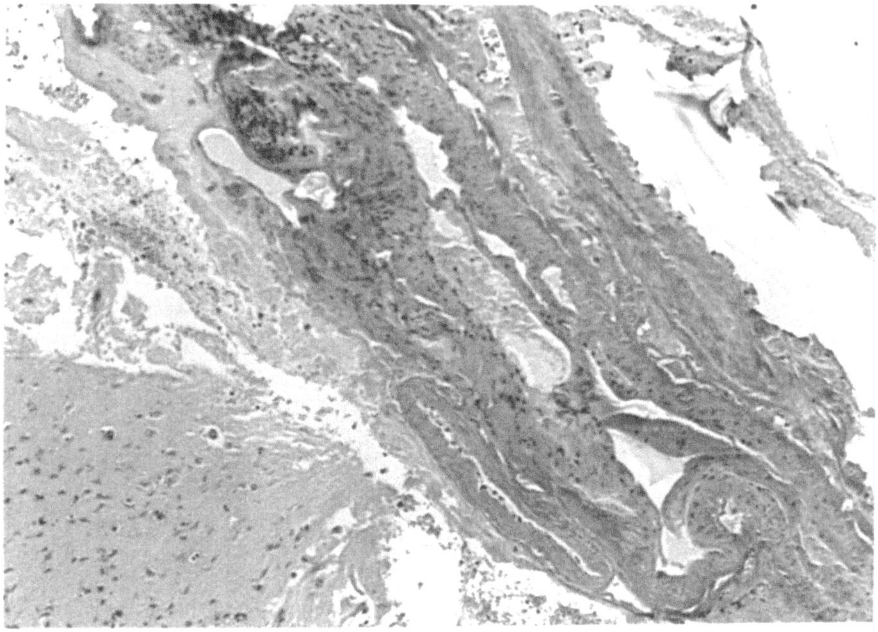

Fig. 12a. Photomicrograph of an AVM after Nd:YAG laser exposure. Brain parenchyme (left lower corner). Leptomeningeal vessels (right upper corner). Coagulation of arteries with swollen vessel wall and narrow lumen. (H. & E. × 40)

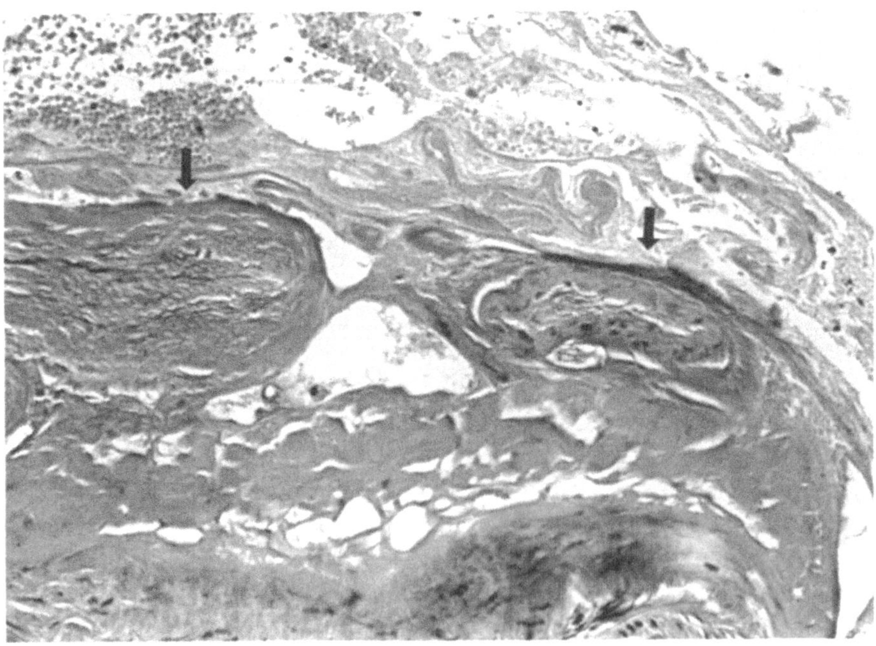

Fig. 12b. Two partially occluded vessels (arrows). Thermal shrinkage of collagen fibers that appear swollen and fused. There is a complete disintegration of the architecture. (H. & E. × 100)

we have not been able to overcome several of the technical problems associated with the obliteration of large, high-flow malformations. These fragile vessels under extraordinarily high pressure contain little contractile tissue and are precariously thin. Precise application of the proper amount of power in the correct site is necessary to obliterate the vessel lumen. If substantial flow is present, the vessel must be occluded in order to facilitate thrombosis. The development of a precise micromanipulator for the surgical microscope has greatly facilitated the precision of application. The spot size can be accurately controlled and the laser beam precisely directed under direct vision without the interposition of vision-obstructing handpieces. Controlled hypotension and tactile occlusion of high-flowing blood in thin-walled vessels will induce thrombosis, but this technique does not provide substantive improvement over coagulation with bipolar current. However, the application of techniques has gained remarkable results in deep-seated malformations of the ventricles, thalamus and brainstem, where direct approach with standard techniques is unusually difficult.

Our experience with twenty malformations of the thalamus documents that the lesions can be totally obliterated and excised without applications of clips, and with minimal need for bipolar current. Previously most of these lesions were considered inoperable, patients suffered recurrent hemorrhage despite radiotherapy and other forms of adjunctive therapy. The laser is most effective in small vessels in which the flow is moderate and some contractile tissue is preserved. Intraoperative and postoperative hemorrhages remain the most serious problems associated with total excision of giant complex high-flow AVMs. Precise hemostasis with the Nd : YAG laser offers the hope of a technical advance in these dangerous lesions which pose a major surgical challenge. We are convinced that the laser will eventually add a new promise of safety and control in the excision of these lesions. The Nd : YAG laser has recently been approved by the U.S. FDA for use as a hemostatic adjunct in excision of vascular malformations. This approval will encourage industrial collaborators to increase their efforts toward development of more effective technical devices for our clinical needs [16, 22].

In our series there was no morbidity or increased neurological deficit attributable to laser techniques. This further supports the evidence in the literature that, when employed judiciously, the laser can be used safely. However, we continue to be wary of use in the posterior fossa, especially in the region of the brainstem, to avoid the disastrous results similar to those reported by Jain [14].

We are encouraged by our experiences with this new technique of coagulation occlusion of arteriovenous malformations. We believe that these techniques have improved operability of many of these lesions. Study of the histologic and physiologic consequences of Nd : YAG energy continues.

Development of new micromanipulators for both CO_2 and Nd : YAG fibers is needed, and endoscopic techniques for Nd : YAG applications hold promise for future development.

References

1. Beck OJ (1980) The use of the Nd: YAG and the CO_2 laser in neurosurgery. Neurosurg Rev 3: 261–266
2. Beck OJ (1984) The use of the Nd: YAG laser in neurosurgery. Neurosurg Rev 7: 151–157
3. Cushing H, Bailey P (1928) Tumors arising from the blood vessels of the brain: angiomatous malformations and hemangioblastomas. Charles C Thomas, Springfield, Illinois, 219 pp
4. Dandy WE (1928) Arteriovenous aneurysms of the brain. Arch Surg 17: 190–243
5. Debrun G, Vinuela F, Fox A, Drake CG (1982) Embolization of cerebral arteriovenous malformations with bucrylate. J Neurosurg 56: 615–627
6. Drake CG (1979) Cerebral arteriovenous malformations: Consideration for and experience with surgical treatment in 166 cases. Clin Neurosurg 26: 145–208
7. Fasano VA (1981) The treatment of vascular malformation of the brain with laser sources. Lasers Surg Med 1: 347–356
8. Fasano VA, Urciuoli R, Ponzio RM (1982) Photocoagulation of cerebral arteriovenous malformations and arterial aneurysms with the neodymium: Yttrium-Aluminum-Garnet or Argon laser: preliminary results in twelve patients. Neurosurgery 11 (6): 754–760
9. Fasano VA (1983) Observations on the use of three laser sources in sequence (CO_2-Argon-Nd: YAG) in neurosurgery. Lasers Surg Med 2: 199–203
10. Forster DMC, Steiner L, Hakanson S (1972) Arteriovenous malformations of the brain. A long-term clinical study. J Neurosurg 37: 562–570
11. Hilal SL, Michelson JW (1975) Therapeutic percutaneous embolization for extraaxial vascular lesions of the head, neck, and spine. J Neurosurg 43: 275–287
12. Jain KK, Gorisch W (1979) Repair of small blood vessels by Neodymium-YAG laser. Surg 85: 684
13. Jain KK (1983) Lasers in neurosurgery: a review. Lasers Surg Med 2: 217–230
14. Jain KK (1985) Complications of use of the Neodymium: Yttrium-Aluminum-Garnet laser in neurosurgery. Neurosurgery 16 (6): 759–762
15. Kamikawa K, Ohnishi T, Hayakawa T, Yoshimine T, Mochida N, Togo T, Takeuchi K (1979) Evaluation of CO_2 laser surgical unit in neurosurgery. In: Laser surgery: Proceedings of the 3rd International Congress for Laser Surgery, Graz, 24–26, 1979, pp 109–114
16. Kelly PF, Alker GJ, Zoll JG (1982) A microstereotactic approach to deep-seated arteriovenous malformations. Surg Neurol 17: 260–262
17. Kjellsberg RN, Hanamura T, Davis KR, Lyons SL, Adams RD (1983) Bragg-peak Proton-beam therapy for arteriovenous malformations of the brain. N Eng J Medicine 309: 269–274

18. Leheta F, Gorisch W (1976) Coagulation of blood vessels by means of argon ion and Nd: YAG laser radiation. In: Kaplan I (ed) Laser surgery I. Jerusalem Academic Press, Jerusalem, pp 178–184
19. Luessenhop AJ, Spence WT (1960) Artificial embolization of cerebral arteries: report of use in a case of arteriovenous malformation. JAMA 172: 1153–1155
20. Luessenhop AJ, Presper JH (1975) Surgical embolization of cerebral arteriovenous malformations through internal carotid and vertebral arteries. J Neurosurg 42: 443–451
21. Michaelson WJ (1979) Natural history and pathophysiology of arteriovenous malformations. Clin Neurosurg 26: 307–313
22. O'Reilly GV, Colucci VM, Astorian DG, Schoene WC, Clarke RH, Hammerschlag SB (March 1982) Transcatheter fiberoptic laser coagulation of blood vessels. Radiology 142: 777–780
23. Parkinson D, Bachers G (1980) Arteriovenous malformations: summary of 100 consecutive supratentorial cases. J Neurosurg 53: 285–299
24. Russell WF, Smith RR (1982) Pellet embolization of central nervous system arteriovenous malformations. In: Smith RR, Haerer AF, Russell WF (eds) Seminars in neurological surgery: vascular malformations and fistulas of the brain. Raven Press, New York, pp 101–117
25. Samson D (1983) Surgical treatment of intracranial arteriovenous malformations. Tex Med 79: 52–59
26. Sang UH, Wilson CB (1975) Surgical treatment of intracranial vascular malformations. West Med J 112: 175–183
27. Stein BM, Wolpert SM (1980) Arteriovenous malformations of the brain. Parts I and II: Current concepts and treatment. Arch Neurol 37: 1–5, 69–75
28. Steiner L, Leksell L, Forster DMC, Greitz T, Backlund EO (1974) Stereotactic radiosurgery in intracranial arterio-venous malformations. Acta Neurochir (Wien) [Suppl] 21: 195–209
29. Tekeuchi J, Handa H, Taki W, Yamagami T (1982) The Nd:YAG laser in neurological surgery. Surg Neurol 18 (2): 140–142
30. Tew JM Jr, Tomsick TA, Lukin RA (1984) Applications of interventional neuroradiology in the treatment of arteriovenous malformation and other vascular disorders of the brain. In: Neurosurgeons, Vol. 4, Proceedings of The Japanese Congress of Neurological Surgeons, Osaka, Japan, pp 331–356, Japan Upjohn Ltd.
31. Tew JM Jr, Tobler WD (1986) Present status of lasers in neurosurgery. In: Symon L *et al* (eds) Advances and technical standards in neurosurgery, Vol 13. Springer, Wien New York, pp 3–36
32. Troupp H, Marttila I, Halonen V (1970) Arteriovenous malformations of the brain: prognosis without operation. Acta Neurochir (Wien) 22: 125–128
33. Wharen RE Jr, Anderson RE, Scheithauer B, Sundt TM Jr (1984) The Nd:YAG laser in neurosurgery. Part 1. Laboratory investigations: dose-related biological response of neural tissue. J Neurosurg 60: 531–539

34. Wharen RE Jr, Anderson RE, Sundt TM Jr (1984) The Nd:YAG laser in neurosurgery. Part 2. Clinical studies: an adjunctive measure for hemostasis in resection of arteriovenous malformations. J Neurosurg 60: 540–547
35. Wolpert SM, Stein BM (1975) Catheter embolization of intracranial arteriovenous malformations as an aid to surgical excision. Neurorad 10: 73–85

Address for correspondence: John M. Tew Jr., M.D., Professor and Chairman, Department of Neurosurgery, University of Cincinnati, College of Medicine, 231 Bethesda Avenue, ML #515, Cincinnati, OH 45267, U.S.A.

Laser Resection of Meningiomas

W. C. Clark and **J. H. Robertson**

Department of Neurosurgery, University of Tennessee, Memphis, Tennessee, U.S.A.

Introduction

The meningioma, being a benign tumor, has been intimately involved in the evolution of neurosurgical technique. Since an operative cure is at least theoretically possible with complete tumor removals, attempts at improvement in intra- and perioperative management have usually resulted in significant advances in neurosurgery.

In 1927, Cushing described his use of the unipolar cautery in the removal of meningiomas of the olfactory groove in his now famous Macewen Memorial Lecture [12]. Since then, further development and evolution of techniques in modern neuroanesthesia, selective cerebral angiography, computed tomography, and the operating microscope have further enhanced the treatment of meningiomas [27]. Within the last decade, lasers have become an established modality in the operative management of intracranial and intraspinal meningiomas because of their ability to allow very precise tissue removal and dissection [3, 33, 39, 41].

The purpose of this chapter is to review the use of the laser in the management of meningiomas of various locations, along with a discussion of the operative techniques as they apply to these tumors in specific locations. Throughout this discussion, the term "laser" is used in the generic sense, and should be interpreted as the carbon dioxide laser unless otherwise noted.

Background

Meningiomas constitute from 13 to 19 percent of all intracranial tumors [5, 13, 20, 35, 43], and about 25 percent of intradural extramedullary spinal tumors [19, 37]. For purposes of discussion, we have divided these tumors according to location (Table 1).

Table 1

I. Intracranial
- A. Supratentorial
  1. Parasagittal and Falx
  2. Convexity
  3. Sphenoidal Ridge
  4. Olfactory groove
  5. Suprasellar
  6. Intraventricular
  7. Other
- B. Infratentorial
  1. Cerebellopontine angle
  2. Tentorium
  3. Clivus
  4. Foramen magnum
  5. Cerebellar convexity

II. Intraspinal
- A. Intradural extramedullary

Intracranial

Supratentorial

About 80 to 90 percent of all intracranial meningiomas are found above the level of the tentorium [27, 35]. Like other extra-axial intracranial tumors, meningiomas cause symptoms and signs primarily due to compression of adjacent brain tissue and cranial nerves, and in this sense there is no specificity to the presenting clinical pattern that indicates the tumor type with certainty [27]. However, the sites of intracranial meningiomas bear a general relationship to the areas of most abundant arachnoid granulations, and the clinical presentation of each tumor is best described in reference to its anatomical location [35].

Parasagittal and Falx

Tumors in these locations constitute about one-third to one-fourth of all intracranial meningiomas [13, 27, 35]. The parasagittal tumor is attached to the superior sagittal sinus and occupies the space between the falx and the convexity dura, indenting the medial surface of the hemisphere [34, 35].

The falx meningioma is not visible on the surface of the brain. It arises from the falx or inferior sagittal sinus, depresses the pericallosal artery and corpus callosum, and indents the medial hemisphere. When the tumor is large it is difficult to differentiate from the parasagittal meningioma. Characteristics that may aid in this differentiation include the tendency for the falcine meningiomas to extend through the falx to become bilateral, and the usual absence of hyperostosis as compared to parasagittal tumors [35].

The importance of dividing the sagittal sinus into thirds for descriptive purposes has long been recognized [13, 26]. Each of these locations tend to present with particular features, and determine 1) the methods used to manage the sinus when it is involved with tumor; 2) the relative importance of the sagittal veins at differing locations; and 3) the presenting and postoperative neurologic manifestations of lesions that occur [26].

The anterior one-third of the sinus extends from the crista galli to the coronal suture. Meningiomas arising in this portion of the sagittal sinus may give little clinical evidence of their location [35]. While they produce headache, seizures, and an insidious deterioration of personality and intellect leading to progressive dementia, a complete discussion of the presenting symptomatology is beyond the scope of this presentation, and one of the excellent reviews of this topic should be consulted [26, 34].

The patient is pretreated with dexamethasone; 4 mg orally every six hours for at least 48 hours prior to surgery, and adequate serum levels of phenytoin are achieved in order to assure anticonvulsant prophylaxis. The patient is positioned in order to minimize intracranial pressure. The head is fixed in the Mayfield pin headrest, rotated no more than about 30 degrees to the contralateral side in order to avoid any obstruction of the neck veins, and elevated about 25 to 30 degrees to promote venous drainage. It is important at this time to be sure the head can be further elevated if excessive bleeding should occur from the sagittal sinus. For tumors in the anterior third of the sinus, the head is slightly extended.

A modified Sutar skin flap is fashioned in order to provide wide exposure of the cranium and allow removal of a free bone flap that crosses the midline and the underlying superior sagittal sinus, and provides wide exposure of the tumor to be resected. Some surgeons prefer a bone flap that does not cross the sagittal sinus because accidental laceration of the sinus wall with a saw may provoke considerable blood loss before the flap can be turned and the bleeding controlled [26]. The authors have not found this to be an insoluble problem, and prefer burr holes one-third over the sinus and two-thirds over dura on either side of the sinus. The sinus is then stripped from the calvarium before the power saw is used, and the cuts over the sinus are made last. The dural flap is hinged medially on the sinus, and is outlined over the tumor as it is lightly palpated with forceps. If the dura is adherent to the tumor, internal decompression may be performed through the involved dura. If not, the free dural flap is hinged medially and secured with dural retraction sutures. Any exposed normal cortex and the sinus are protected by wet cottonoid strips or Gelfoam. The operating microscope and laser are brought into position. Suction-irrigation is used during the internal decompression to further decrease the spread of heat to surrounding tissues. It is usually necessary to rely upon bipolar coagulation for hemostasis because the laser cannot deal with vessels much larger than 1.0 to 1.5 mm in diameter. The internal decompression is performed using about 60 watts of continuous, defocused laser power. If there is calcification within the tumor, larger amounts of power may be necessary (80–100 watt range). By using larger amounts of power and abundant irrigation the authors have found that tumor tissue is evaporated more cleanly with less carbonization, and the internal decompression proceeds much more rapidly.

Once the tumor bulk is reduced, the capsule becomes more pliable, and may be folded inward on itself. The capsule is pulled medially and either cut off in segments with miroscissors or evaporated piecemeal with the laser. The gap created by retracting the tumor capsule inward allows separation of the arachnoid from it. Any arteries attached to the capsule are carefully dissected free until it can be confirmed that they are actually running into and supplying the tumor. Only then are they coagulated or clipped. The capsule is then removed bit by bit until all that remains is the portion that is attached to the falx or sinus. In cases where the attachment is in the anterior one-third of the sinus, the sinus may be sacrificed without risking serious sequelae.

The middle third of the sagittal sinus extends from the coronal suture to the lambdoid suture, and incorporates the motor and sensory areas and the Rolandic veins. Middle third tumors classically present with focal epilepsy, usually commencing in the foot. Progressive hemiparesis or sensory loss usually starts in the leg, but as the tumor extends laterally the arm becomes involved and may become the most severely affected limb [35]. Preoperative digital subtraction angiography is of critical importance in the assessment of the patency of the sinus.

Preoperative steroids and anticonvulsants are used in the routine fashion. Positioning is the same as for the anterior third tumors except that the head is slightly flexed.

Either a biparietal or temporal skin flap is fashioned, with the midline and at least an additional 2 to 3 cm beyond exposed. A free bone flap that crosses the midline is removed as described earlier. The opening of the dura and the initial debulking of the tumor is carried out in the same manner as anterior third tumors. Once the capsule is removed except for the portion that is attached to the sinus, the next step is determined by the extent of sinus involvement.

A classification of the level of sinus involvement has been proposed by Bonnal [7], and a variety of authors have proposed different methods of repairing the sinus after removal of a parasagittal meningioma [6, 21, 22,

26, 31]. Kapp and Schmidek [36] discuss a variety of these repairs in detail. If the tumor is attached only to the wall of the sinus, the site of attachment may be coagulated with the bipolar forceps or carefully evaporated with low wattage laser energy until a clean, shiny dural surface is obtained. In cases where a portion of the tumor lies in a dural reflection of the lateral wall or roof of the sinus, the external layer of the sinus is usually all that is involved, and this may be removed leaving the internal layer and the venous sinus intact. If the tumor invades only the lateral angle of the sinus, the sinus is opened, the tumor evaporated or removed with microsurgical technique and the sinus then closed with nonabsorbable suture. In cases where the preoperative arteriogram has shown the sinus is completely occluded, preservation of the neighboring cortical veins is even more critical because it is assumed that they constitute at least a portion of the collateral circulation that has prevented venous infarction and serious preoperative neurologic deficits. A clinical decision must then be made whether to attempt resection of the sinus. In the event the operator decides to attack the occluded sinus, various authors have recommended use of a total autogenous vein graft repair [6, 22], pericranial graft repair [26], lyophilized dura [21], or other means of reconstituting the sagittal sinus.

Once the tumor has been removed, the sinus appropriately managed, and adequate hemostasis secured, the dural defect is repaired, the bone flap replaced, and the scalp closed in the customary fashion.

The posterior third of the sagittal sinus extends from the level of the lambdoid suture to the torcular Herophili. The major signs and symptoms of meningiomas in this area are slowly progressive visual field defects, and often present as a homonymous hemianopia or visual hallucinations of the occipital lobe type. Because of the contiguous lateral sinuses at the level of the torcular, preoperative digital subtraction angiography becomes even more important in the assessment of the venous drainage consisting of demonstration of the patency of the sinus, dominance of one or the other lateral sinuses, and the presence of venous collaterals.

Patients with posterior third tumors are operated in the prone position, with the head fixed in a pin headrest and rotated toward the side of the tumor to provide maximal exposure. Most venous drainage in this area is from the lateral and inferior aspects of the occipital lobe into the lateral sinus, the superior petrosal sinus, and the adjacent tentorial veins. The scalp and free bone flaps are fashioned in the customary manner. The approach down to the tumor and its exenteration is completed in the usual fashion followed by retraction of the tumor capsule medially, away from the cortex and its evaporation or excision so as to leave only the basal attachment. At this point the extent of sinus involvement is assessed. If there is only adherence to the sinus the tumor is removed and the dura coagulated with low power, defocused laser energy. On the other hand, sinus involvement presents a much more difficult problem. The sinus in this area is not easily obliterated with finger pressure to control hemorrhage while excising a portion of the sinus without causing severe venous distention and edema. As a result, the surgeon usually has to be content with coagulation of the tumor on the sinus wall unless the sinus is completely occluded. If the sinus is occluded and there are no large sagittal veins on either side, the sinus may be excised as previously described.

Convexity

These tumors are free of attachment to the dural sinuses and tend to be concentrated in the region adjacent to the coronal suture [18]. Various reports have placed their incidence varying from about 13 to 34 percent of intracranial meningiomas [13, 27, 35].

The vascular supply to these tumors is often via the middle meningeal artery or one of its branches. Once this supply is interrupted and the dura opened, the pulsations of the brain often tend to aid in the delivery of the tumor mass.

The techniques of surgical removal vary little from those previously described, and primarily consist of an internal decompression using high wattage, defocused, continuous laser energy. Once the capsule becomes pliable it is retracted inward on itself and evaporated or excised. Decompression and capsule removal continue in a stepwise fashion until total removal is achieved.

Depending on the size of these tumors and their level of adherence to brain, alternatives for removal include the cutting loop and the ultrasonic surgical aspirator.

Sphenoidal Ridge

The sphenoid ridge is the boundary between the anterior and middle fossae. From the anterior clinoid process medially, the lesser wing of the sphenoid bone comprises the inner two thirds of the ridge and the greater wing makes up the outer third, which terminates laterally at the pterion. Meningiomas in this area may encroach upon the anterior fossa, middle fossa, sylvian fissure, orbit, and cavernous sinus [35]. These tumors usually make up about 18 to 20 percent of series of meningiomas [13, 27, 35].

The presentation of these tumors is primarily determined by the site of origin of these tumors along the sphenoid wing [27], which Cushing divided into the inner or clinoidal, middle or alar, and outer or pterional portions [13].

Tumors that arise at the anterior clinoid process typically cause unilateral loss of vision and primary optic atrophy as a result of optic nerve compression. If the optic tract or chiasm is also involved, there may be an incongruous homonymous type field defect, but the visual loss is more pronounced in the ipsilateral eye [23]. Cranial nerve involvement may occur at the superior orbital fissure, and most often is the abducens nerve [27]. Hypesthesia in the distribution of the first division of the trigeminal nerve may also occur [1], and is most often associated with cavernous sinus involvement [27].

Medial sphenoid ridge meningiomas continue to present a challenge to the neurosurgeon. Traditionally, the major problems associated with the removal of these tumors occur as a result of the tumor's intimate attachments to major neurovascular structures deep within the cranial cavity, and these structures still occasionally cause difficulties. However, the limits of resectability of these tumors must now be considered carotid artery involvement (rather than encasement) and significant cavernous sinus invasion and involvement without occlusion, which is fortunately rare.

Scalp and bone flaps are reflected so as to expose the sphenoid ridge and allow removal of the squamous portion of the temporal bone to the floor of the middle fossa. The sphenoid wing is removed with rongeurs and a high speed drill to the level of the orbito-meningeal artery. This vessel is coagulated and severed, providing a flat, basal, tangential approach to the supraclinoid carotid artery and the tumor. Cerebrospinal fluid is removed from the lateral chiasmatic, basal sylvian, and carotid cisterns in order to facilitate brain relaxation and subsequent retraction. An attempt is made to identify the limits of the mass in all directions, and identify its relationships to the optic nerve, chiasm, and tracts, the internal carotid artery and the tentorial edge at the level of the posterior clinoid process. This is often precluded by the bulk of the tumor, and initial internal decompression is required to allow sufficient mobilization of the tumor capsule so that the full extent of the mass and its attachments may be appreciated.

Internal decompression of the tumor is carried out using a high power (about 60 watt), defocused laser beam in a continuous fashion. Continuous suction-irrigation is used in order to minimize the spread of heat to surrounding tissues. This internal decompression must be carried out with great care since it is possible to encounter large arterial trunks that have been encased by the tumor mass. It is occasionally of value to identify the vessel outside the mass, and then divide and evaporate the tumor as the plane around the vessel is followed into the mass. When working very close to the vessel, other structures to be preserved, or deep in the tumor mass adjacent to the distal tumor capsule one has the option of using the laser in a pulsed rather than continuous fashion, or if available, the superpulse mode in order to decrease the spread of heat to adjacent tissue. Excessive displacement of the tumor should not be attempted until the carotid artery and optic nerve are identified. In most instances the optic nerve will be displaced ventrally and be markedly compressed. Arachnoid strands can be divided and the nerve separated from tumor capsule. The carotid artery is also often markedly displaced, but an adequate plane of dissection is often present and allows the vessel to be freed from the tumor. The careful internal decompression and extracapsular dissection is continued until sufficient tumor mass has been removed that the mass may be retracted superiorly with the frontal lobe in order to expose and inspect the free edge of the tentorium. Once these attachments have been severed, attention is returned to the major vascular structures. Further internal decompression is carried out as necessary to allow mobilization of the capsule sufficient to expose the proximal anterior and middle cerebral arteries and any association they have with the tumor.

With the major vascular structures identified and separated from the tumor, internal decompression and piecemeal capsule removal is continued in order to minimize traction on the brain. After removal of the tumor is completed, a low power defocused laser beam is used to thoroughly coagulate the entire dural surface along the sphenoid, with special attention directed toward the area of previous attachment. Hemostasis is achieved using the bipolar cautery and Gelfoam soaked in thrombin. The wound is then closed in a customary fashion.

Meningiomas of the middle and lateral thirds of the sphenoid wing generally reach a larger size before making their presence known, either with the hyperostosing changes of bone or with seizures or symptoms of increased intracranial pressure [35]. The surgical removal of these lateral tumors is not as hazardous as with those tumors located on the medial sphenoid wing. A standard pterional craniotomy is used in order to expose the tumor. The boundaries are identified and outlined with Gelfoam strips. Once again, an internal decompression of the mass is performed as previously described. Once the capsule becomes empty and can be manipulated, it can either be evaporated with the laser or excised with microsurgical instruments.

Once the tumor mass has been eradicated, the dural and bony attachments of the lesion must be removed. This usually requires coagulation or evaporation of the dura in the middle fossa and along the greater and lesser wings of the sphenoidal bone. This is accomplished with a "painting" of the area using a low energy, defocused laser beam in a continuous fashion. In areas where the bone of the base of the skull has been infiltrated by meningioma, the laser may again be used to evaporate or devitalize most of this invasion. This represents a significant advancement in technique, since recurrence of these tumors is often due to this invading portion that previously either was not recognized or could not be easily or cosmetically removed [7].

Olfactory Groove

These tumors arise over the lamina cribrosa of the ethmoid, indenting the frontal lobes inferiorly, and commonly they are bilateral [2, 35]. Anosmia, the earliest manifestation of the tumor, is nearly always present, although the patient rarely complains of it [35]. In most reported series of intracranial meningiomas, they constitute somewhere between five and thirteen percent of these tumors.

While David and Askenasy attemped to organize and classify these tumors into anterior, middle, or posterior groups based on their relation to the floor of the anterior fossa [14], it has been our experience and is the feeling of others [2] that the large size of these tumors at the time of presentation usually precludes any clinically meaningful classification along these lines. It is important to recognize that these tumors are often bilateral [35], and this should be taken into account when planning the operative approach to these tumors.

With high resolution computed tomography it is now usually possible to differentiate those tumors with bilateral extension. In the event the tumor is bilateral, a modified Sutar skin flap and bifrontal free bone flap is utilized. Otherwise, the approach is through a standard frontal craniotomy. There is usually remarkable edema associated with these tumors. As a result, an effort is made to apply only minimal retraction forces in achieving the exposure.

An internal decompression of the tumor is accomplished as described earlier, but extreme care must be taken because of the typical displacement of the anterior cerebral arteries, optic nerves, and hypothalamus that occurs when these tumors are large.

Suprasellar

This group of meningiomas consists only of those tumors arising from the tuberculum sellae, planum sphenoidale, diaphragma sellae, and/or the anterior clinoid processes. These tumors usually comprise only about ten percent of all intracranial meningiomas [11, 13, 16, 27, 35, 40], and their size, close proximity to the optic apparatus and major vessels, and their vascularity and usually hard consistency have made them a considerable technical challenge for the neurosurgeon [40].

The most common presenting complaint is usually visual loss, with the characteristic picture often described as a bitemporal hemianopia that occurs as a result of upward pressure on the optic chiasm. This classic presentation is used to attempt to differentiate these tumors from those of the olfactory groove, since they tend to press on the optic nerves from above and cause scotomata that are usually easily differentiated from field cuts. Less commonly, one sees headache, mental changes, and seizures as the initial complaint with these tumors [40].

The tumor may be approached through a frontal, bifrontal, or pterional approach. On most cases the authors prefer the pterional approach from the nondominant side. The craniotomy is performed in the customary fashion, with the sphenoid ridge being removed extradurally down to the level of the orbitomeningeal artery. The dura is then opened close and parallel to the floor of the anterior and middle fossae. A relaxing incision is also made in the dura in the direction of the sylvian fissure in order to facilitate dural retraction and exposure. The frontal and temporal lobes are gently retracted in order to allow cerebrospinal fluid drainage from the chiasmatic, carotid, and sylvian cisterns and increase the amount of cerebral relaxation. Self-retaining retractors are placed and the operating microscope and laser brought into the operating field. The olfactory tract is then identified, and at the point where it crosses the sphenoid wing it usually serves as a reliable landmark of the entry of the optic nerve into the optic canal.

Once the tumor and the extent of its relations to the surrounding neurovascular structures have been identified, tumor removal is begun. The initial step is an internal decompression performed as described earlier, using continuous suction-irrigation in order to protect the surrounding structures. Once an adequate debulking is completed, the capsule becomes sufficiently thin and easily mobile. At this point the capsule is either dissected with microsurgical technique or evaporated in order to remove it from the optic nerves, chiasm, and blood vessels. The dural attachment is then heavily coagulated or evaporated using a low power, defocused laser beam in a continuous fashion.

Hemostasis is achieved with bipolar coagulation, the defocused laser beam, and Gelfoam soaked in thrombin, and the wound is closed in a customary fashion.

Intraventricular

Meningiomas of the ventricular system are relatively rare, and have been found to comprise from 0.5 to 2.0 percent of various series of intracranial meningiomas [17, 24, 27, 35]. There would appear to be a tendency for these tumors to occur more often in the lateral ventricles [17, 24], on the left more than the right [4, 17, 24], although they have also been reported in both the third and fourth ventricles [30, 42]. Most of the lateral ventricular meningiomas originate in the posterior portion of the lateral ventricle, especially in the region of the trigone, and receive their blood supply primarily from the choroidal arteries [4, 29]. These tumors arise from the arachnoid of the choroid plexus, and may be fibrous, meningotheliomatous or mixed [25].

As with other intraventricular tumors, the intraventricular meningioma continues to present a significant technical challenge. There are a variety of approaches to these tumors described in the literature [15, 27, 38], but the basic concepts include planning to avoid functional areas, using minimal retraction of the hemisphere, and controlling the main blood supply of the tumor as early as possible [38]. Since the principles of laser removal of these lesions is the same, regardless of their location in the ventricle, the approach and techniques of removal of the trigonal tumors will be described, recognizing that its adaptation for the removal of tumors in the body, frontal and temporal horns will require certain changes as a result of the subtle differences in each of these locations.

Most surgeons recognize that one single approach to the trigonal tumor is not the best for all tumors. The authors prefer the posterior parietal approach just above the lambdoidal suture, but also recognize certain circumstances will arise whan a lateral temporal parietal, middle temporal gyrus, superior parietal occipital, or some combination of these incisions will provide a safer and better exposure of a given tumor. The relative merits of each of these approaches for trigonal tumors have been discussed, and the reader is referred to that discussion for more detailed information [38]. Also, the authors prefer an incision rather than a cortical resection because of the increased functional loss that may accompany resection in this area.

The cortical incision may be performed in a standard fashion using bipolar coagulation, or, alternatively using a low wattage defocused laser beam. It is very important to recognize that one of the keys to the successful removal of these tumors is that one should not attempt to visualize the borders of the tumor by retracting on the white matter. Rather, an intracapsular removal should be accomplished using a defocused laser beam, with the borders of the tumor seen only when sufficient decompression has been carried out to allow the tumor capsule to be folded in on itself. The surgeon must be prepared to deal with blood loss, since the choroidal artery tributaries supplying the tumor cannot be controlled until some of the tumor mass has been removed. The use of the laser provides the capability to remove these masses from within the ventricle through a small exposure, with improved hemostasis, and without applying mechanical forces to the choroid plexus that can result in undetected bleeding at a distance from the operative field.

Other

The remaining one to two percent of supratentorial meningiomas are comprised of those found in the middle fossa, the orbit, and miscellaneous rare ectopic sites such as the paranasal sinuses, the neck, and the temporal bone. The principles of the use of the laser for removal of these tumors is consistent with techniques already described.

Infratentorial

Ten to twenty percent of all intracranial meningiomas are found below the level of the tentorium [8, 27, 35]. A study by Castellano and Riggiero classified the meningiomas into five groups, and an adaptation of that classification will be utilized here [8]. Like their supratentorial counterparts, meningiomas of the posterior fossa cause symptoms and signs primarily due to compression of adjacent brain tissue and cranial nerves, and, again there is not specificity to the presenting clinical pattern that would differentiate a meningioma from a neuroma with certainty. Meningiomas of the posterior fossa are thus classified according to their site of attachment in the cerebellopontine angle, the tentorium cerebelli, the clivus, the foramen magnum, and tumors of the cerebellar convexity.

Cerebellopontine Angle (CPA)

Meningiomas arising along the posterior surface of the petrous bone adjacent to the porus acusticus in the cerebellopontine angle comprise 40 percent of the posterior fossa meningiomas. They cause symptoms similar to those associated with acoustic neuromas and cannot be distinguished with assurance on clinical findings alone [32]. The more recent use of high resolution computed tomography and magnetic resonance imaging have greatly aided this differential, with meningiomas tending to show less erosion of the porus and being closely applied and contiguous with the dura overlying the petrous ridge. Other signs that aid in this differential are the tendencies for meningiomas to show less marked elevations of the cerebrospinal fluid protein [35] and more common involvement of the fifth and lower four cranial nerves [32].

The surgical approach most often utilized for these tumors is the standard suboccipital craniectomy. A variety of incisions and skin flaps have been described for this approach, but the authors prefer a paramedian, lazy S-shaped incision centered over the mastoid groove. The craniectomy is extended to the transverse sinus superiorly, the lateral sinus laterally, and the foramen magnum inferiorly. Sufficient bone is removed medially to obtain good visualization of the tumor. The dura is then opened with a Y-shaped incision and the cerebellar cisterns entered in order to obtain release of cerebrospinal fluid and cerebellar relaxation. The cerebellum is gently retracted with a self-retaining retractor in order to expose the tumor. The authors have generally found it unnecessary to resect any of the cerebellum in achieving adequate tumor exposure.

The lower cranial nerves are first identified and dissected from the tumor capsule. They are protected with Gelfoam or wet cottonoids. The tumor capsule is evaporated with a defocused laser beam, and an intracapsular decompression is carried out as previously described. The amount of power used is based upon the texture of the tumor and the surgeon's personal preferences. However, the authors have found that tough, fibrous tumors and those with calcification generally require from 40 to 80 watts of defocused CO_2 laser power in order to aggressively proceed with the debulking of the tumor mass. A word of caution is in order since the amount of power and duration of application of the beam must be decreased as the opposite capsular wall is approached. Further, important structures such as the anterior inferior cerebellar artery (AICA) and the facial nerve may be displaced in almost any direction, or completely concealed by the tumor. This displacement is usually far more variable than that seen with acoustic neuromas [32], even though these structures are usually extracapsular. The authors have recently begun routine electromyographic monitoring and direct facial nerve stimulation in an effort to preserve facial function, and have been impressed with this technique.

Once the internal decompression is completed, a thin capsule is all that remains. The capsular wall is carefully involuted and contiguous structures are dissected off the capsule. Depending on how pliable the capsule may be, it is either removed with microdissection or evaporated with the laser. If the laser is used, the operator should protect the structures that have been dissected free with Gelfoam or wet cottonoids.

Tentorium Cerebelli

Meningiomas of the tentorium comprise 30 percent of posterior fossa meningiomas. These tumors commonly invade the venous sinuses and may extend above as well as below the tentorium [35]. Castellano and Riggiero analyzed Olivecrona's series of meningiomas and concluded that tentorial meningiomas are best removed from above whether they present on the superior or inferior surface of the tentorium [8]. It is now probably more reasonable to conclude that the approach to these tumors should be individualized, the approach being determined by the location of the main mass of the tumor. Small tumors growing downward into the posterior fossa are approached by a subtemporal-occipital approach with tentorial splitting. Tentorial meningiomas with a large element in the posterior fossa or those involving the posterior petrous ridge are better approached by a combined supra- and infratentorial approach [32].

The skin incision is begun about six centimeters above the ear and again is extended in an S-shaped fashion centered over the mastoid groove. The inferior portion of the incision in the neck is outlined, prepped, and draped so the incision can be extended as far inferiorly as needed to achieve adequate exposure. Initially a temporo-occipital free bone flap is elevated, with additional bone removed as necessary to allow flat, direct access over the petrous bone to the tentorium. The dural flap is based on the lateral sinus in order to avoid interrupting bridging veins before they are well exposed. Occipital bridging veins are coagulated with bipolar forceps and cut well away from the dura and lateral sinus. The occipital lobe is gently retracted a little further to expose the tentorium and the incisura. If no tumor is visible, gentle palpation with microinstruments will often reveal the location of the tumor arising from the underside of the tentorium. The tentorium is then incised well away from the tumor attachment and the lateral sinus, with hemostasis usually possible with bipolar coagulation. It is occasionally necessary to use metallic clips, but they are not routinely required. The tentorium is then further opened using a low power defocused laser beam and bipolar coagulation. Extension of the opening is carried forward to the incisura, and posteriorly to the transverse sinus. Care must be taken not to injure the trochlear nerve at the incisura or the transverse sinus at the posterior aspect of the incision. This then results in an opening into the cerebellopontine angle from above, and may suffice for tumor removal provided the tumor is not large. If there is a large extension into the cerebellopontine angle or the tumor arises from the posterior surface of the petrous ridge, the incision is extended and a suboccipital craniectomy performed as described earlier in the discussion of CPA meningiomas.

Once the tumor is exposed, an internal decompression is performed with the defocused laser beam, with great care exercised not to place the brainstem or the cranial nerves under compression or traction by manipulating the tumor. Even though there is a higher incidence of meningiomas of the angioblastic variety in this region, the authors have not yet encountered one where hemostasis could not be achieved alternately using the CO_2 laser and bipolar coagulation. Because of this increased incidence of angioblastic meningiomas, the neodymium: yttrium-aluminum-garnet (Nd: YAG) laser would at least have theoretical advantages due to its superior hemostatic abilities [3]. However, great caution would be in order as a result of its decreased precision and widespread heating effects in the proximity of the brainstem and cranial nerves [19].

Once the internal decompression is complete, the capsule should be sufficiently pliable to allow it to be folded inward on itself, allowing the operator to avoid retracting on the brainstem or cranial nerves. The residual pieces of capsule may be removed piecemeal by microdissection or laser evaporation.

Clivus

Clivus meningiomas account for ten percent of all posterior fossa meningiomas [35]. In a series reported before the use of microsurgical techniques, 16 of 29 patients with clivus meningioma died within two weeks of operation, with the one year survival after diagnosis being 25 percent [9]. A more recent series has been reported that utilized an inventive section of the nondominant venous sinus at the junction of the lateral and transverse sinuses and the tentorium to expose the clivus and achieve complete removals in many [28]. The clivus meningiomas may be approached in a variety of ways. These include the transclival approach through either a transseptal or sublabial incision, a transtentorial approach through the middle fossa, or approaches through the posterior fossa.

Regardless of the approach selected, removal is carried out as with other meningiomas, by first performing a careful internal decompression with the defocused laser beam. The removal is then completed as previously described. Meningiomas of the clivus are distinctly uncommon, and the clinician must be ever aware of the possibility of this lesion in order to make the diagnosis [9].

Foramen magnum

Foramen magnum meningiomas account for eight percent of the posterior fossa meningiomas [35]. The site of dural attachment is most often along the anterior rim of the foramen magnum [27]. While the diagnosis is most often confirmed with high resolution computed tomography, preoperative arteriography delineates the relationship of the tumor to the vertebral artery, and magnetic resonance imaging shows the direction and degree of brainstem and spinal cord displacement.

A paramedian incision extends from the level of the inion to about C-3. A wide craniectomy is performed in order to completely expose the inferior portion of the cerebellar hemisphere, with the incision extended to expose the spinal cord down to about C_2—C_3. The dura is opened with a Y-shaped incision and retracted with sutures. Cerebellar relaxation is obtained by removal of cerebrospinal fluid from the cistern. The entrance of the vertebral artery into the intrathecal space is identified by finding the uppermost dentate ligament. The spinal accessory nerve is identified and protected with a wet cottonoid. Extracapsular dissection is performed as much as possible without putting traction on surrounding structures in order to allow Gelfoam or wet cottonoids to be placed around the capsule to: 1) prevent laser damage to the surrounding tissues; and 2) decrease the amount of subarachnoid blood that may cause a postoperative aseptic meningitis. The tumor is then enucleated using a defocused laser beam as described previously. As the capsule is collapsed inward it is removed with microdissection or evaporated with the laser.

Cerebellar Convexity

These tumors constitute eight percent of the posterior fossa series, and are usually found in close proximity to the transverse sinus, and may be attached to the torcular Herophili [35]. If tumor involvement occurs on the nondominant side or the sinus is occluded by tumor, the sinus usually can be resected if necessary. The full extent of any venous sinus involvement is routinely ascertained with preoperative arteriography.

The tumors are approached through a routine suboccipital craniectomy whose superior border is the transverse sinus. The dura is opened and reflected upward on the sinus. Great care should be taken in this maneuver in case the tumor is adherent to the dura or the sinus. If the tumor is merely adherent to the sinus, it is removed as described earlier for parasagittal tumors. If it is adherent only to the dura, the dura is removed along with the tumor mass, and a free graft of pericranium placed. In cases where the tumor is extremely adherent to both the dura and the tumor bed in the cerebellar hemisphere, an internal decompression is carried out as described previously in order to allow the tumor and its dural attachment to be evaporated and folded inward on itself. This facilitates both exposure and microdissection of the capsule from the cerebellum. A careful watertight closure of the dura combined with meticulous hemostasis and closure of the wound completes the procedure.

Intraspinal

Intradural-Extramedullary

Meningiomas of the spinal canal are small, rounded tumors, and only become symptomatic late in their course as a result of very slow growth. They are most often found in the thoracic region (81%), with far fewer located in the cervical (16%) or lumbar (3%) areas [19, 37]. Most subdural meningiomas are attached to the anterolateral wall of the dura, with only a few attached to the posterior wall [19]. When a meningioma is attached to the posterior or posterolateral wall of the dura, it is often possible to incise the dura around the tumor's attachment, achieve dural hemostasis with bipolar coagulation, and simply lift the tumor from its bed in the spinal cord. If the tumor capsule is adherent to the cord or there is difficulty in identifying the dural attachment, the laser may be used to internally decompress the tumor mass and allow retraction only on the tumor itself rather than the spinal cord. This compression is performed in a fairly standard fashion using about 40 watts of defocused laser power in a continuous fashion. After the majority of the tumor is decompressed the power is decreased and the laser is used in pulsed bursts to prevent the spread of heat through the remaining tumor tissue and capsule to the cord below. Evoked potential monitoring can be used as a guide of when to stop the dissection, add irrigation to help dissipate heat, etc. The authors also advocate the use of continuous suction-irrigation in order to aid the control of heat generated during the laser dissection.

In anterolateral meningiomas, the spinal cord appears displaced backwards and to the opposite side with the nerve roots and dentate ligaments stretched. The dentate ligaments are secured with a small suture for retraction and then sectioned. An effort should be made to preserve any nerve roots, but this is often very difficult to do. The spinal cord is then rotated delicately by retracting slightly on the dentate ligaments and the tumor is exposed. When the tumor is large and the space available insufficient, the operator may remove one or more articular facets on the side of the tumor in order to improve the exposure. The spinal cord and nerve roots are then protected with cotton pledgets soaked in physiological saline solution. The laser is then used to effect an internal decompression of the tumor mass. This allows the operator to retract on the capsule to begin separating it from the cord without applying major distracting forces to the spinal cord itself. Once the capsule has been emptied, the volume of the mass is usually sufficiently decreased to allow removal of the entire capsule along with its dural attachment.

In meningiomas located anterior to the spinal cord, the tumor should be approached using an anterior or anteriolateral approach as described by several authors. Once adequate operative exposure is achieved the internal decompression and subsequent tumor removal is accomplished as with the other intraspinal meningiomas. After removal of the meningioma, a free graft of bank dura mater, synthetic dural substitute, or fascia lata is applied to the ventral surface of the cord and folded back on the dorsal surface like a hammock in an attempt to prevent a cerebrospinal fluid leak [19].

The indications for the use of the carbon dioxide surgical laser attached to the operating microscope in the removal of intraspinal meningiomas is based upon 1) the location of the tumor in relationship to the spinal cord, 2) the vascularity of the tumor, and 3) the consistency of the mass. The laser is used to internally decompress the tumor mass without applying major distracting forces to the spinal cord. This general technique of cavitation is accomplished with a defocused beam of carbon dioxide laser in a range of 30–60 watts. The level of energy used in the decompression is related to the size of the tumor mass and its consistency. Mobilization of the tumor capsule is accomplished with the bipolar cautery and microsurgical dissection. Dura involved by tumor is removed by the defocused carbon dioxide laser beam.

References

1. Ausman JI, French LA, Baker AB (1974) Intracranial neoplasms. In: Baker AB, Baker LH (eds) Clinical neurology. Harper and Row, Hagerstown, MD, pp 1–103
2. Bakay L, Cares HL (1972) Olfactory meningiomas. Acta Neurochir (Wien) 26: 1–12

3. Beck OJ (1980) The use of the Nd:YAG and the CO_2 laser in neurosurgery. Neurosurg Rev 3: 261–266
4. Bernasconi V, Cabrini GP (1967) Radiological features of tumors of the lateral ventricles. Acta Neurochir (Wien) 17: 290–310
5. Boldrey E (1971) The meningiomas. In: Minckler S (ed) Pathology of the nervous system. McGraw-Hill, New York, pp 2125–2144
6. Bonnal J, Brotchi J, Stevenaert A *et al* (1971) L'ablation de la portion intra-sinusale des meningiomes parasagittaux rolandiques, suirie de plastie du sinus longitudinal superieur. Neurochirurgie 17: 341–354
7. Bonnal J, Brotchi J (1978) Surgery of the sagittal sinus in parasagittal meningiomas. J Neurosurg 48: 935–945
8. Castellano F, Riggiero G (1953) Meningiomas of the posterior fossa. Acta Radiol (Stockholm) [Suppl] 104: 1–177
9. Cherington M, Schneck SA (1966) Clivus meningiomas. Neurology 16: 86–92
10. Clark WC, Robertson JH, Gardner G (1984) Selective absorption and control of thermal effects. A comparison of the systems used in otology and neurotology. Otolaryngol Head Neck Surg 92: 73–79
11. Clark WC, Acker JD, Robertson JH *et al* (1986) Reformatted sagittal images in the differential diagnosis of meningiomas and pituitary adenomas with suprasellar extension. Neurosurg 18: 555–558
12. Cushing HW (1969) The meningiomas arising from the olfactory groove and their removal by the aid of electrosurgery. In: Matson DD, German WJ (eds) Harvey Cushing: selected papers on neurosurgery. Yale University Press, New Haven, Conn, pp 246–273
13. Cushing HW, Eisenhardt L (1938) The meningiomas: Their classification, regional behavior, life history, and surgical end results. Ch C Thomas, Springfield, Ill
14. David M, Askenasy H (1937) Les meningiomes olfactifs. Rev Neurol 68: 489–531
15. De LaTorre E, Alexander E, Davis CH *et al* (1963) Tumors of the lateral ventricles of the brain. Report of eight cases with suggestions for clinical management. J Neurosurg 20: 461–470
16. Finn JE, Mount LA (1974) Meningiomas of the tuberculum seelae and planum sphenoidale. Arch Ophthalmol 92: 23–27
17. Fornari M, Savoiardo M, Morello G (1981) Meningiomas of the lateral ventricles. J Neurosurg 54: 64–74
18. Giombini S, Solero CL, Morello G (1984) Late outcome of operations for supratentorial convexity meningiomas. Surg Neurol 22: 588–594
19. Guidetti B (1974) Removal of extramedullary benign spinal cord tumors. In: Krayenbühl H *et al* (eds) Advances and technical standards in neurosurgery, Vol 1. Springer, Wien New York, pp 173–197
20. Grant FC (1954) A clinical experience with meningiomas of the brain. J Neurosurg 11: 479–487
21. Hartmann K, Klug W (1975) Recidivation and possibilities of surgery in meningiomas of the middle and posterior third of the longitudinal sinus. In: Klug W, Brock M, Klinger M *et al* (eds) Advances in neurosurgery, me-

ningiomas, multiple sclerosis, forensic problems in neurosurgery. Springer, Berlin Heidelberg New York, pp 100–107
22. Kapp JP, Gielchinsky I, Deardourff SL (1977) Operative techniques for management of lesions involving the dural venous sinuses. Surg Neurol 7: 339–342
23. Kearns TP, Wagener HP (1953) Ophthalmologic diagnosis of meningiomas of the sphenoid ridge. Am J Med Sci 226: 221–228
24. Kobayashi S, Okazaki H, MacCarty CS (1971) Intraventricular meningiomas, Mayo Clin Proc 46: 735–741
25. Landenheim JC (1963) Choroid plexus meningiomas of the lateral ventricle. Ch C Thomas, Springfield, Ill
26. Logue V (1975) Parasagittal meningiomas. In: Krayenbühl H *et al* (eds) Advances and technical standards in neurosurgery, Vol 2. Springer, Wien New York, pp 171–198
27. MacCarty CS, Piepgras DG, Ebersold MJ (1982) Meningeal tumors of the brain. In: Yeomans J (ed) Neurological surgery. Saunders, Philadelphia, pp 2936–2966
28. Malis LI (1985) Tumors of the skull base. A presentation at the American Association of Neurological Surgeons, Denver, CO, April
29. Mani RL, Hedgcock MW, Mass Si *et al* (1978) Radiographic diagnosis of meningiomas of the lateral ventricle. J Neurosurg 49: 249–255
30. Markwalder TM, Markwalder RV, Markwalder HM (1979) Meningioma of the anterior part of the third ventricle. J Neurosurg 50: 233–235
31. Masuzawa H, Tamura A, Sano K (1977) Recurrence of parasagittal meningioma and sinus-plasty using falx. Sixth International Congress of Neurological Surgery, Sao Paulo, Brazil, June, 1977 [Abstract], Excerpta Medica, International Congress Series No 418, p 121
32. Maxwell RE, Chou SN (1982) Posterior fossa meningiomas. In: Schmidek HH, Sweet WH (eds) Operative neurosurgical techniqes. Grune and Stratton, New York, pp 517–533
33. Nishiura I, Handa H, Yamashita J *et al* (1981) Successful removal of a huge falcotentorial meningioma by use of the laser. Surg Neurol 16: 380–385
34. Northfield DWC (1973) The surgery of the central nervous system. A textbook for postgraduate students. Blackwell, Oxford
35. Quest DO (1978) Meningiomas. An update. Neurosurg 3: 219–225
36. Schmidek HH, Kapp JP (1984) Traumatic and neoplastic involvement of the cerebral venous system. In: Kapp JP, Schmidek HH (eds) The cerebral venous system and it's disorders. Grune and Stratton, New York, pp 581–596
37. Simeone FA (1975) Intraspinal neoplasms. In: Rothman RA, Simeone FA (eds) The Spine. Saunders, Philadelphia, pp 823–836
38. Spencer DD, Collins WF (1982) Surgical management of lateral intraventricular tumors. In: Schmidek HH, Sweek WH (eds) Operative neurosurgical techniques. Grune and Stratton, New York, pp 561–574
39. Strait TA, Robertson JH, Clark WC (1982) Use of the carbon dioxide laser in the operative management of intracranial meningiomas. A report of twenty cases. Neurosurg 10: 464–467

40. Symon L, Rosenstein J (1984) Surgical management of suprasellar meningioma. J Neurosurg 61: 633–641
41. Takizawa T, Yamazaki T, Miura N *et al* (1980) Laser surgery of basal, orbital, and ventricular meningiomas which are difficult to extirpate by conventional methods. Neurol Med Chu (Tokyo) 20: 719–737
42. Tsuboi K, Nose T, Maki Y (1983) Meningioma of the fourth ventricle. Neurosurg 13: 163–166
43. Zülch KJ (1957) Brain tumors. Springer, Wien

Address for correspondence: W. Craig Clark, M.D., Ph.D., Assistant Professor, Department of Neurosurgery, University of Tennessee, Memphis, 956 Court Avenue, Coleman Building, Room A-202, Memphis, TN 38163, U.S.A.

Tumours on and in the Pons and Medulla oblongata

P. W. Ascher

Universitätsklinik für Neurochirurgie, Karl-Franzens-Universität Graz, Austria

Introduction

Until the early 1970s [106] tumours of the pons and the medulla oblongata were considered inoperable. These tumours occur mainly in childhood and adolescence, but also in adults. The differential diagnosis, especially in the adult patient, includes thrombosis, vascular lesions and AV-malformations with (or without) hemorrhage.

Before Dandy [76] reported the first successful removal of an encapsulated intrapontine hematoma in 1945, all surgical forays into this area had been followed by catastrophe. Obrador [110] performed the same operation in one patient in 1970.

Interest in tumours of the brain stem began to increase in the 1960s. Large autopsical studies [75, 107, 112] showed results similar to those of Buckley [74], who reviewed 1737 cases operated on by Cushing. Walker and Hoppel [114] as well as Lassman and Arjona [105] reviewed all brain tumours in children and found about 10% to be inoperable lesions of the brain stem. The relatively recent investigative interest in these tumours is due in part to the difficulty of their diagnosis [1, 77, 102, 103, 109, 111, 113, 115]. Before the advent of the CT and NMR, clinicians had to rely on such unspecific symptoms as headache, papilloedema, separation of the cranial sutures and the plethora of neurological deficits resulting from the close anatomical neighborhood of cranial nerve nuclei and ascending and descending tracts. Histologically we find many types of tumour, many of them benign. Surgery was reserved for intrapontine hematomas; patients with other tumours underwent high-voltage treatment and chemotherapy.

In 1975 we introduced the laser as a surgical instrument at the Universitätsklinik für Neurochirurgie in Graz, Austria [2–73, 78–101]. In contrast to earlier neurosurgical attempts [Rosomoff, Fox, Stellar], we used a specially adapted CO_2 laser. The above mentioned authors had worked with other types of lasers and a CO_2 laser designed for general surgery. Poor results and, in our opinion, wrong indications led to a virtual ban of the laser in neurosurgery. Frustrated by the inoperability of benign but deadly tumours and challenged to prove the usefulness of the laser, we found brain stem tumours to be the ideal indication for this new instrument. For the first time it was possible to divide nervous tissue without mechanical or electrical irritation of the surroundings. This non-touch technique is a consequence of the physical characteristics of the CO_2 laser. Corresponding to its wavelength of 10.6 μ, this laser is absorbed in water and, therefore, human tissue to nearly 100%. A focused laser beam permits a clean, clear cut (0.02 mm); a defocused laser vaporizes tumour tissue. This represents a completely new surgical modality.

Against our initial hopes, the CO_2 laser is not ideal for coagulation; only under the microscope is one able to coagulate small vessels. This led us to introduce the Nd:YAG laser, a crystal laser with a wavelength of 1.06 μ, for coagulation. Because of the shorter wavelength, the absorption of the Nd:YAG laser is lower and its penetration deeper. These characteristics, often useful elsewhere, prohibit the routine use of this instrument in and around the brainstem.

Advantages of the Laser

Theoretical considerations of laser applicability were subsequently confirmed experimentally. We were able to show that thermal damage to the

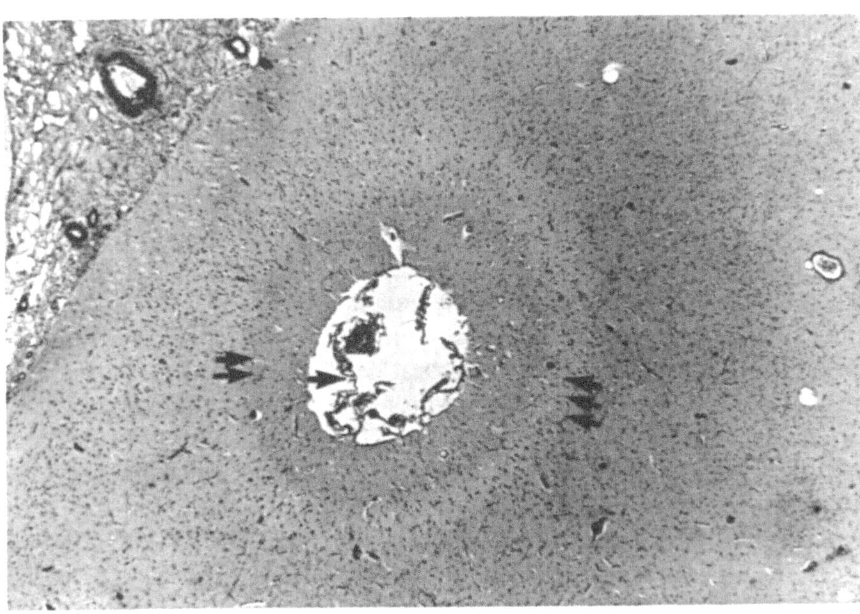

Fig. 1. Laser lesion (30 w, 1 sec., c.w.). Charred mantle (one arrow); coagulation zone (two arrows); zone of edema (three arrows)

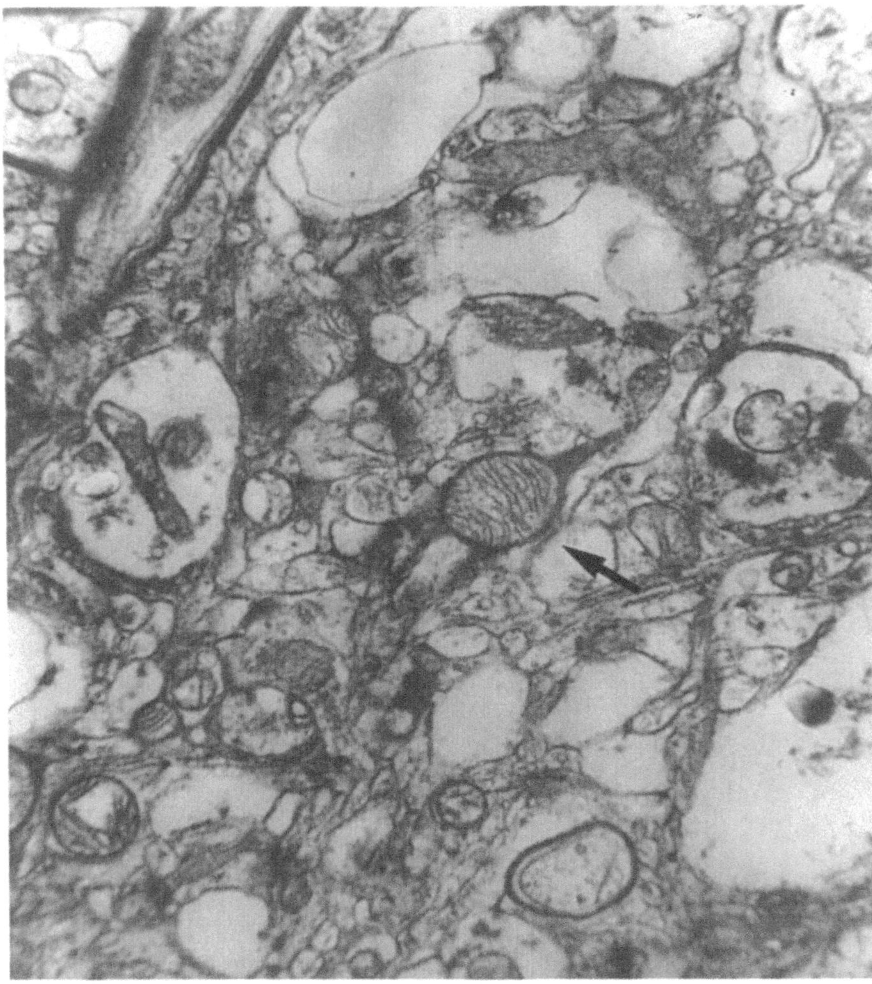

Fig. 2. Intact mitochondria (arrow) in the zone of edema (× 30,000)

surrounding tissue is limited to 0.2 mm. The vaporization crater is bounded by a charred mantle (5–10 A) and a zone of coagulation (0.15 mm) which is continuous to a zone of reversible edema (0.15 mm) (Fig. 1). Electron microscope studies (× 30,000) (Fig. 2) show submicroscopic structures like mitochondria totally intact at a distance of 0.2 mm. Long-term histologic follow-up documents formation of clean glial scars. To evaluate the cutting plane, we compared the cutting surfaces of different surgical modalities (cold knife, electric needle, CO_2 laser, Nd : YAG laser). Scanning electron microscope images (× 2,200) reveal the superiority of the CO_2 laser, which produces an absolutely smooth, even and sealed-like surface (Fig. 3).

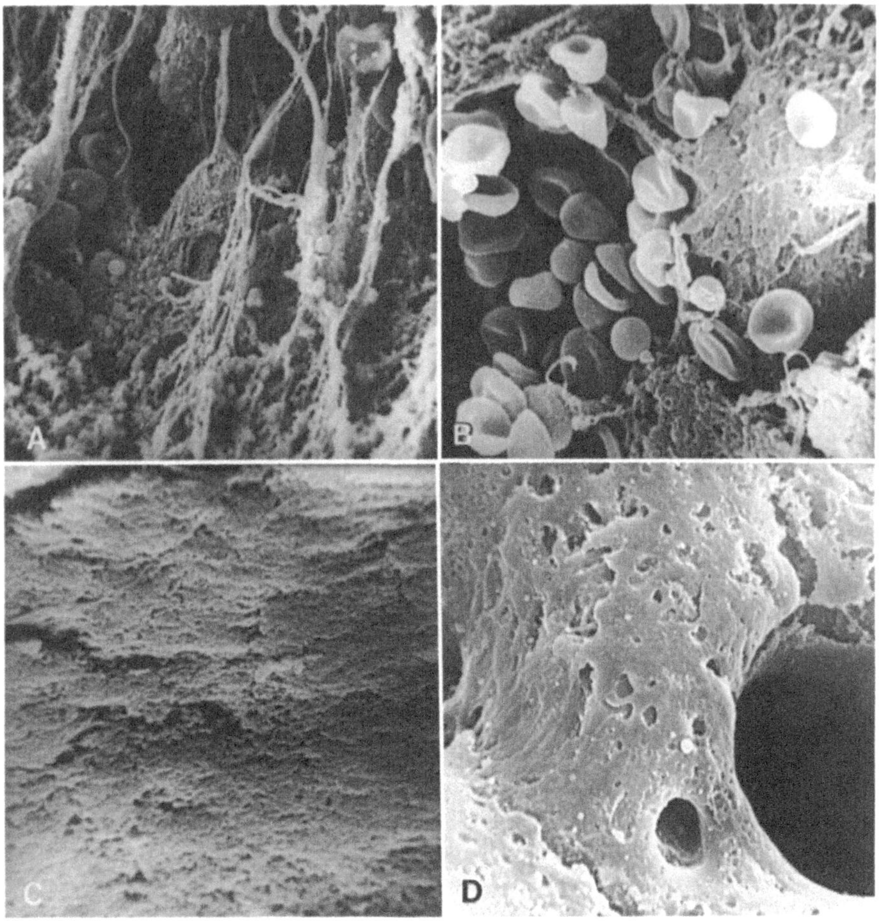

Fig. 3. Brain tissue, SEM (× 2,200). A) Cut with a cold knife, B) cut with electric needle, C) cut with CO_2 laser, smooth and sealed surfaces caused by superficial heat absorption, D) cut with Nd:YAG, porous surface produced by deep heat absorption

The penetration of the CO_2 laser is determined by the output energy of the laser (watts), its speed of movement and the degree of focus.

In addition to offering precision and non-touch technique, the laser does not obstruct the narrow and deep operating field of the neurosurgeon. Apart from these major advantages, the CO_2 laser decreases blood loss, causes no contamination, leads to minimal scar formation, does not interfere with electrical monitoring of the patient and allows absolute protection of the adjacent tissue by a film of water or wet gauze. For the patient, reduced edema formation shortens postoperative intensive care and hospitalization. To the surgeon the laser represents an elegant method

Table 1. *Operations carried out with the laser at the Universitätsklinik für Neurochirurgie in Graz (June 28, 1976–July 18, 1987)*

Glioblastomas	248
Astrocytomas	183
Meningiomas	145
Metastasis	123
Others	129
Tumors of the pineal gland	10
Adenomas of the pituitary gland	12
Plexus papillomas	5
Medulloblastomas	18
Craniopharyngeomas	6
Acoustic nerve tumours	10
Angiomas & AVM	29
Ependymomas	23
Tumors of the corpus callosum	1
Others, not classified tumors	29
Other intracranial processes	49
Hemorrhagic cysts	18
Arachnoidal cysts	8
Abscesses	9
Tuberculomas	1
Procedures on epileptics	13
Operations on the spinal cord	79
Extramedullary tumors	39
Intramedullary tumors	20
Functional surgery	6
Nucleus vaporization	24
Operations on peripheral nerves	33
Operations on vessels	4
Craniostenosis	7
Miscellaneous	56
Total operations with lasers	1,115
Operations with CO_2 laser	846
Operations with Nd:YAG laser	267
Operations with the AR laser	2

Carried out at the University Clinic of Neurosurgery Graz, Austria

Table 2. *Laser operations on and in the brain stem (July 1976–July 1987)*

Histology	
Astrocytoma	28
Ependymoma	19
Medulloblastoma	13
Meningioma	13
Pineal tumours	6
Angiomatous lesions	15
Spongioblastoma	3
Cholesteatoma	1
Papilloma	3
Lipoma	1
Epidermoid cyst	1
Sarcoma	1
Glioblastoma	1
Neurinoma	1
Not classified	3
Metastases	4
Total	113

(Department of Neurosurgery, University of Graz, Austria)

Table 3. *Laser surgery in or close to pons and medulla oblongata (1976–1987)*

Site	Pons	Rhomb-encephalon	Cerebellum pons	Cerebellum rhomb.	
Medulloblastoma	1	5	—	7	13
Astrocytoma	2	3	1	1	7
Spongioblastoma	1	—	—	1	2
Ependymoma	1	7	1	2	11
Papilloma	1	1	—	—	2
Sarcoma	1	—	—	—	1
Angioma	3	1	1	—	5
Metastasis	—	—	2	1	3
Not classified	—	1	1	1	3
Total	7	18	6	13	44

(Department of Neurosurgery, University of Graz, Austria)

Table 4. *Brain stem tumors of childhood (1976–1986)*. Compared to Walker's larger series (115), we found equal mortality in both locations

	Number	Location	Type of surgery		Type of laser application	Op. mortality
Medulloblastomas	10	rhomboid fossa	excision resection	6 4	vaporization	1 edema
Astrocytomas	3	medulla oblongata	resection excision	2 1	preparation	1 pneumonia
Ependymomas	1	medulla oblongata	excision	1	vaporization	0

(Department of Neurosurgery, University of Graz, Austria)

which reduces operating time and tightens control. Because a laser is light, one can operate around an obstruction with the use of metal mirrors. At the time of publication, the first CO_2 laser fibre optics should be available.

Methods and Results

Since 28 July 1976, we have performed move than 1,000 operations with the laser (Table 1). These represent 7% of all operations at the Universitätsklinik für Neurochirurgie in Graz. 113 procedures were on and in the brain stem (Table 2); 47 of these were in the pons or medulla oblongata (Table 3).

The primary mortality of operations on the pons and medulla oblongata is of paramount interest. In our series of 47 patients (14 children, Table 4) there were no intraoperative deaths. We lost 7 patients in the immediate postoperative period (3 weeks). 3 died of cerebral edema, 2 of pulmonary embolism, 1 of pneumonitis and 1 of intestinal hemorrhage (before the introduction of pirenzepine). This postoperative mortality of 15% is low considering the desperate situation of these patients.

7 additional patients died in the first 8 months after surgery: 6 of tumour malignancy (glioma, medulloblastoma etc.) and one after improper use of the wrong laser. In this case, an 8 year old boy with a highly vascularized invasive medulloblastoma was operated on with a Nd:YAG laser. In an effort to achieve hemostasis "safely", wattage was reduced and radiation time prolonged. This led to uniform volume coagulation (protein denaturation) of healthy pontine tissue. The patient remained apallic until death 8 months later; no autopsy was performed.

The remaining 27 patients underwent rehabilitation therapy of variable duration. 10 patients can be considered cured, the remainder showed improvement.

Case Presentation

We present 4 representative patients to illustrate history, neurological status, diagnostic studies, surgical procedures and follow-up.

Case 1: M.C., a 29 year old female, was admitted to another hospital with speech disturbances, intermittent diplopia and hemiparesis. 2 years earlier amyelotic malignant melanomas had been excised from the left thigh and fourth left toe. A CT scan revealed extensive metastasis to the pons (Fig. 4). The patient was transferred to the Universitätsklinik für Neurochirurgie in Graz.

Operation: After endotracheal intubation and general anesthesia and with the patient in a sitting position, an occipital trepanation was performed. After incision of the dura mater, the cisterna magna was opened by splitting the arachnoidea. Inspection showed a bloodily suffused area 10 mm in front of the calamus scriptorius exactly in the midline. The sulcus

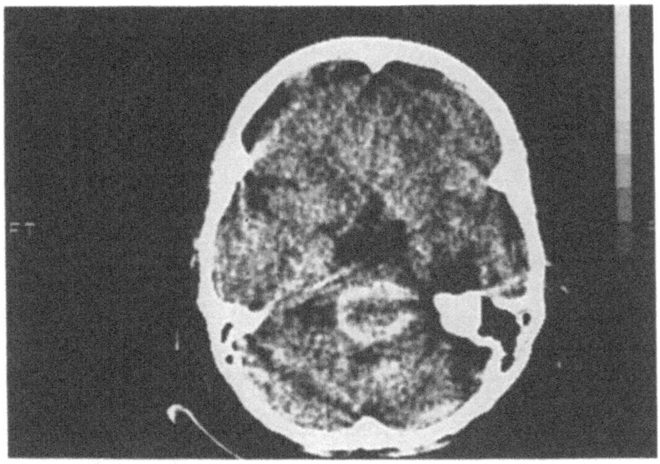

Fig. 4. CT scan shows spherical lesion of the lower brain stem

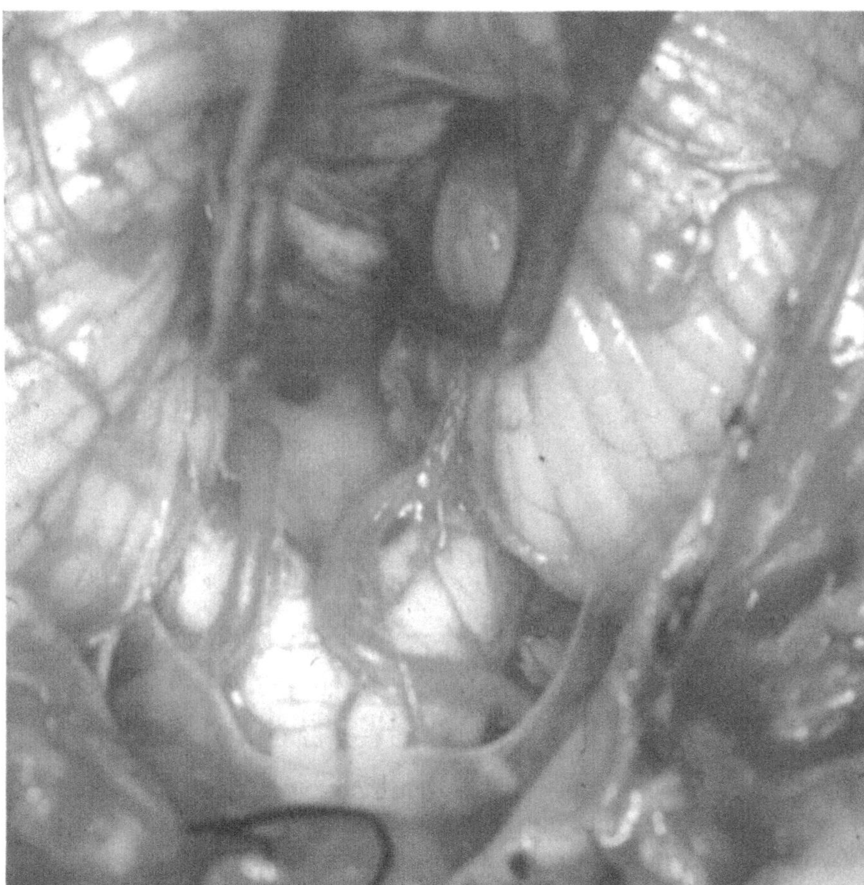

Fig. 5. Intraoperative picture shows exposure of the brain stem and a bloodily suffused area exactly in the midline

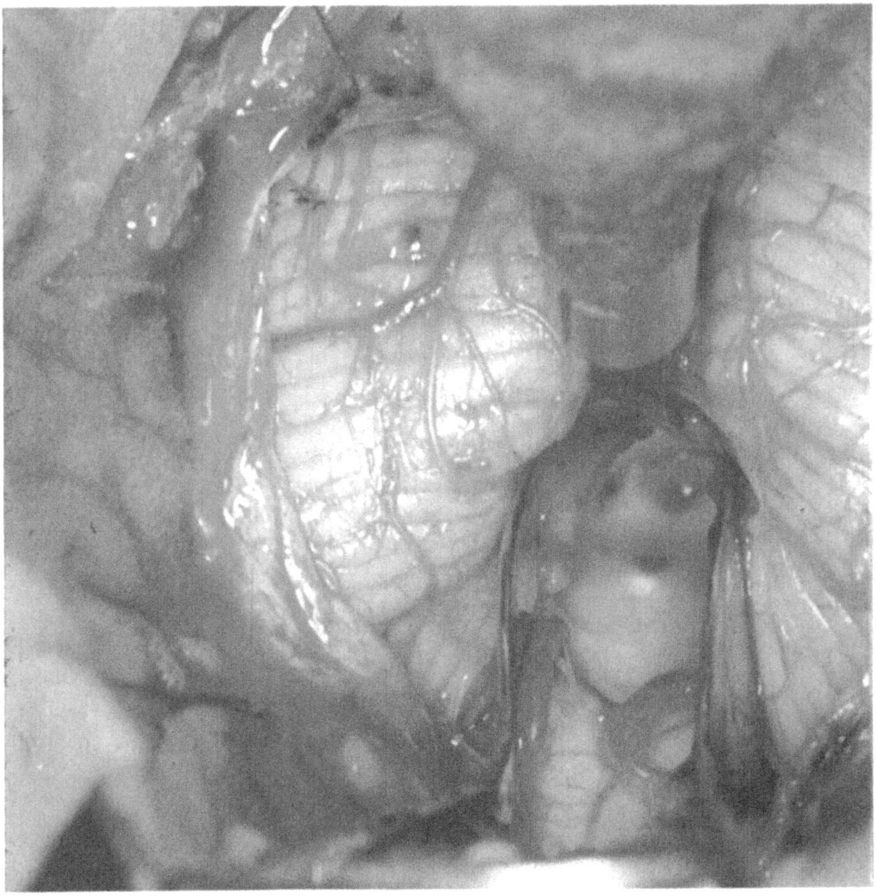

Fig. 6. Operative situation just before use of the laser

medianus was split in the midline with the CO_2 laser (output energy 10 watts, focused, continuous wave (cw), Figs. 5 and 6). After incision of the ventricular ependyma and subependymal glia (1 mm), a dark, bloody cyst was opened (Fig. 7). Material was excised for histology which confirmed the diagnosis of melanoma metastasis. The cyst was emptied by suction and laser vaporization (15 w, defocused, cw). The procedure was completed in the usual way.

The sutures were removed on the seventh postoperative day, the patient was mobilized on the ninth day. Four weeks after surgery the patient was released and referred to the original hospital for postoperative irradiation. The patient died 7 months after surgery.

Case 2: D.B., a 46 year-old-female, was being followed at another hospital because of a 2-year history of unspecific ocular muscle complaints.

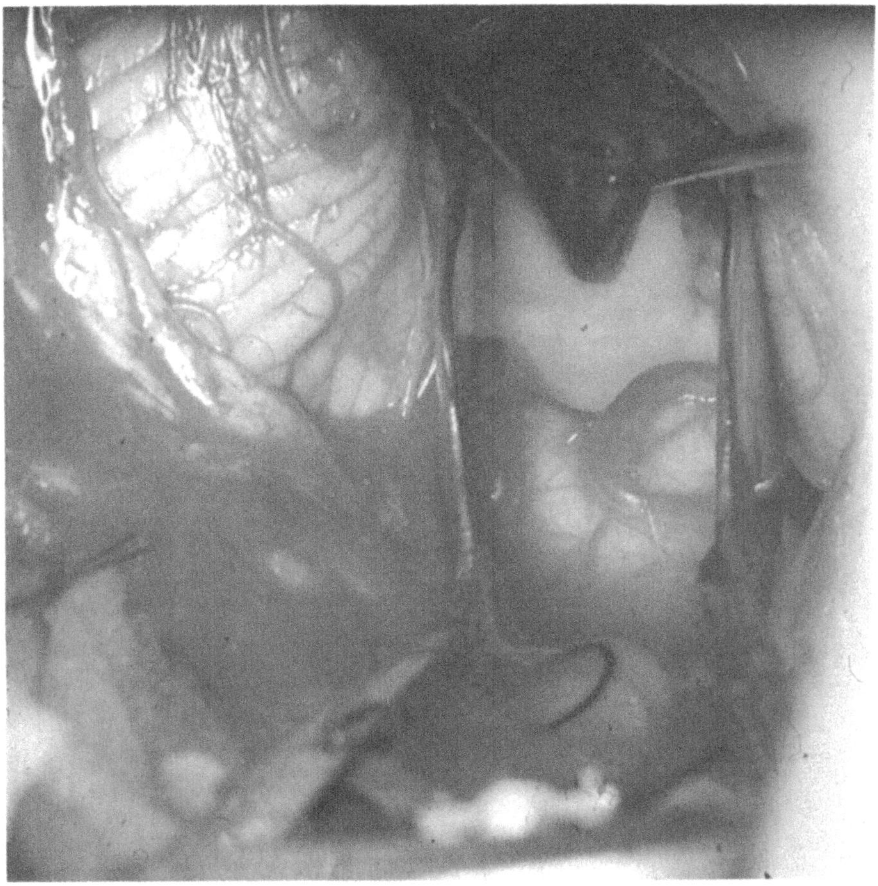

Fig. 7. Operative situation after opening of the tumour cyst

A CT scan showed an expansive process on the floor of the fourth ventricle (Fig. 8). The patient was transferred to our hospital and at admission complained of facial and oral dysesthesia. Examination showed nystagmus on rightward gaze. A new CT scan (Fig. 9) and ventriculography revealed a tumour in the lower mesencephalon (Fig. 10).

Operation: After endotracheal intubation and general anesthesia and with the patient in a sitting position, an occipital trepanation was performed. After incision of the dura mater, the cisterna magna was opened by splitting the arachnoidea. The lower vermis was split with CO_2 laser (10 w, minimally defocused, cw). Inspection of the fossa rhomboidea revealed a bloody, black lesion, 1 cm in diameter, just in front of the aqueductus Sylvii. The lesion had a central perforation out of which old blood oozed (Fig. 11). The perforation was enlarged with a CO_2 laser (10 w,

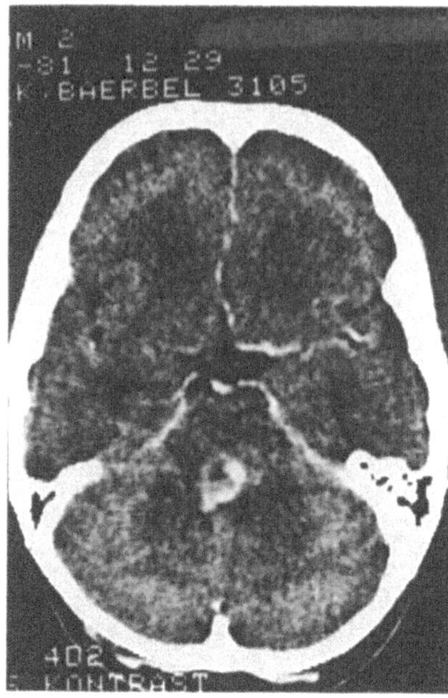

Fig. 8. CT scans shows hyperdense lesion on the floor of the fourth ventricle

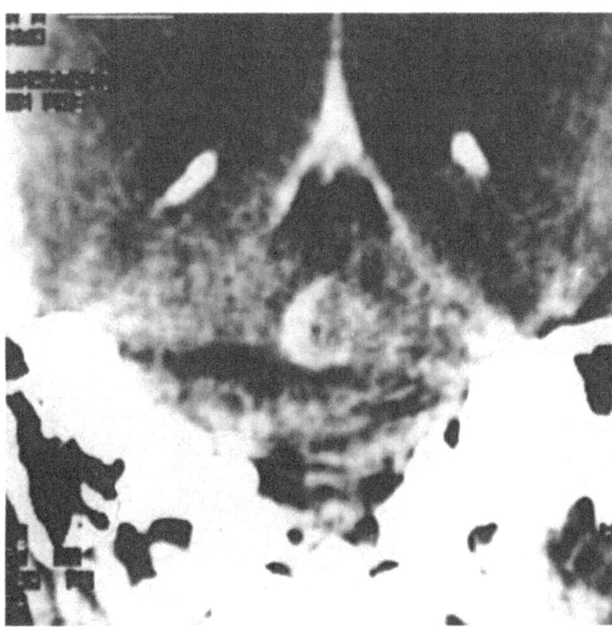

Fig. 9. CT scan confirms previous findings (closer look)

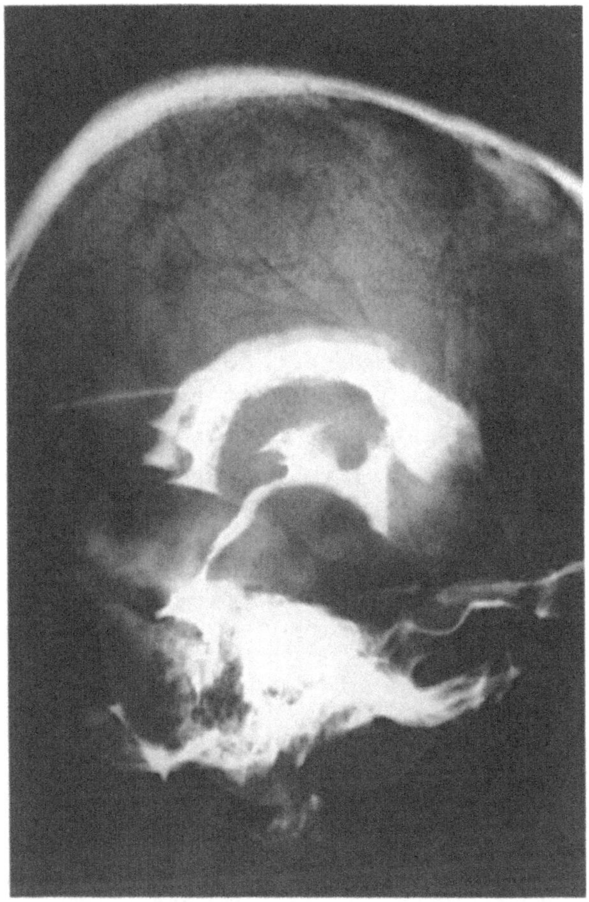

Fig. 10. Positive ventriculography confirms tumour of the pons

focused, repeated single shots of 0.10 s) (Fig. 12). Suction exposed a cavity with a tough coating; histology showed a microadenoma with a hemorrhagic cyst. Hemostasis was attained with the CO_2 laser (10 w, defocused, repeated single shots). The procedure was completed in the usual fashion.

The sutures were removed on the seventh postoperative day, the patient was mobilized on the tenth and released home on the 24th day. The patient, a housewife, achieved full rehabilitation.

Case 3: S. H.-P., a 35-year old male, was beset by strong occipital headaches, which radiated to the temporal regions, 4 months prior to admission. A diagnosis of cervical migraine attacks was made at another hospital. A CT scan was interpreted as negative (Fig. 13). At presentation at our hospital, the patient held his head fixed posteriorly and to the right. There was paresis of the sixth cranial nerve and diplopia. Renewed inter-

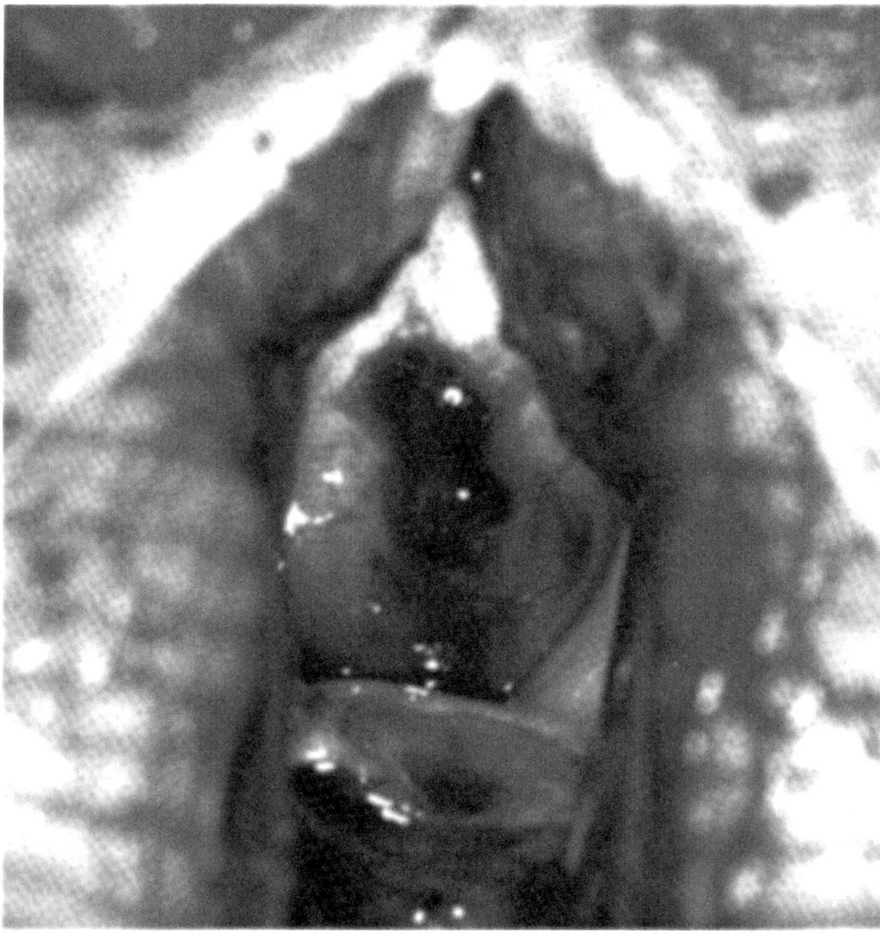

Fig. 11. Intraoperative situation, blood-oozing lesion on the floor of the fourth ventricle

pretation of the original CT scan led to the suspiscion of an expansive process in the lower vermis.

Operation: After endotracheal intubation and general anesthesia and with the patient in a sitting position, an occipital craniotomy was performed. The dura mater, which was bulging outward, was split revealing a blue-grey tumour of variable consistency which was resected in a piecemeal fashion with a CO_2 laser (10 w, foxused, cw). A cystic portion of the tumour was emptied of chocolate-brown fluid by suction. Further resection revealed a part of the tumour affixed to the left side of the obex. During dissection with the laser (10 w, focused, single shots) and forceps, the vital functions twice fell critically. At this point to avoid mechanical irritation

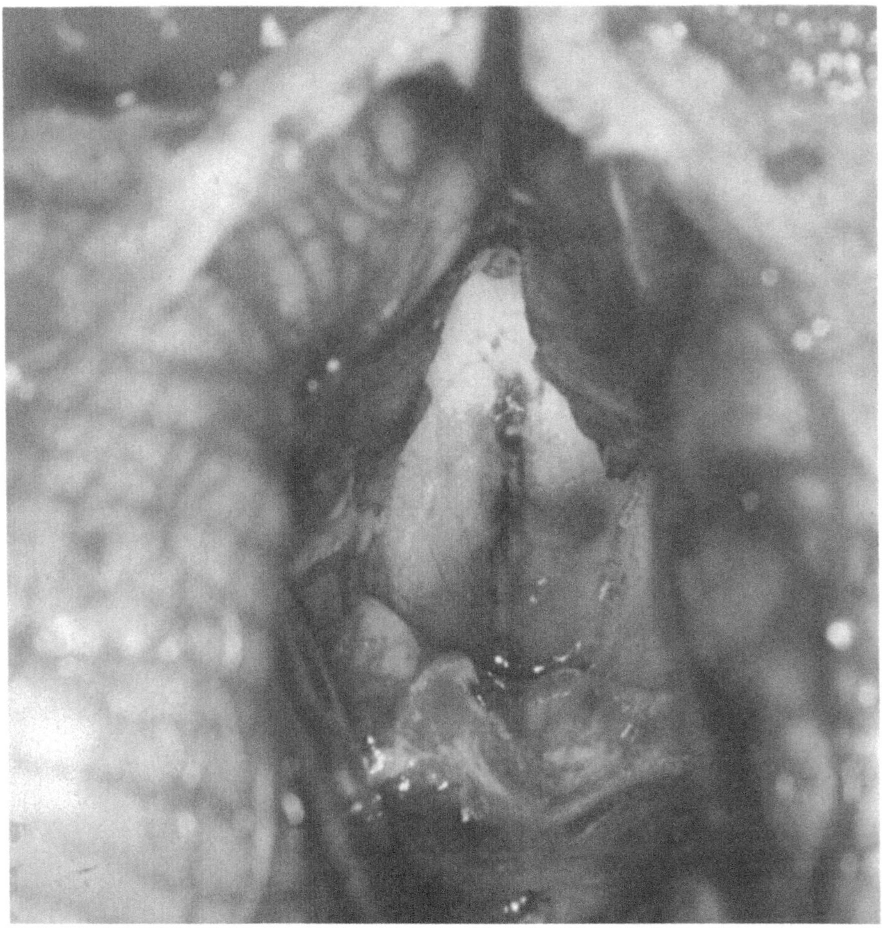

Fig. 12. Opening and emptying of cyst with CO_2 laser and suction

of the brain stem, we changed the technique and vaporized the tumour tissue with the CO_2 laser (15 w, defocused, cw) layer by layer. Histology showed a grade I ependymoma, a diagnosis that forced us to attempt eradication. After vaporization of the remaining tumour, the brain stem was left with an oval identation (2.0 cm × 1 cm × 0.6 cm). The procedure was completed in the usual way.

Two hours postoperatively the patient awoke without neurological deficits. Three days later he developed difficulty swallowing and a leakage of cerebrospinal fluid. 8 days after surgery the patient was reoperated and, because of aspiration pneumonitis, underwent tracheotomy 2 days later. He was in intensive care for 40 days postoperatively, then slowly mobilized and transferred to a rehabilitation center one week later. After 2 months

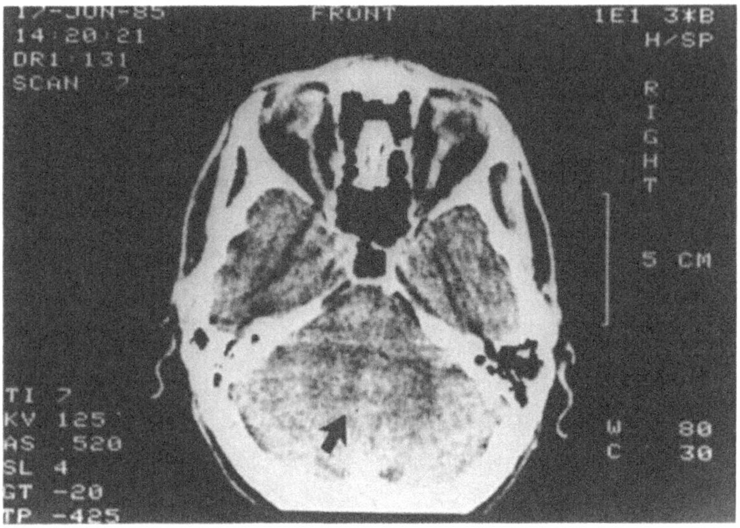

Fig. 13. CT scan shows isodense mass lesion in fourth ventricle (arrow), originally interpreted as negative

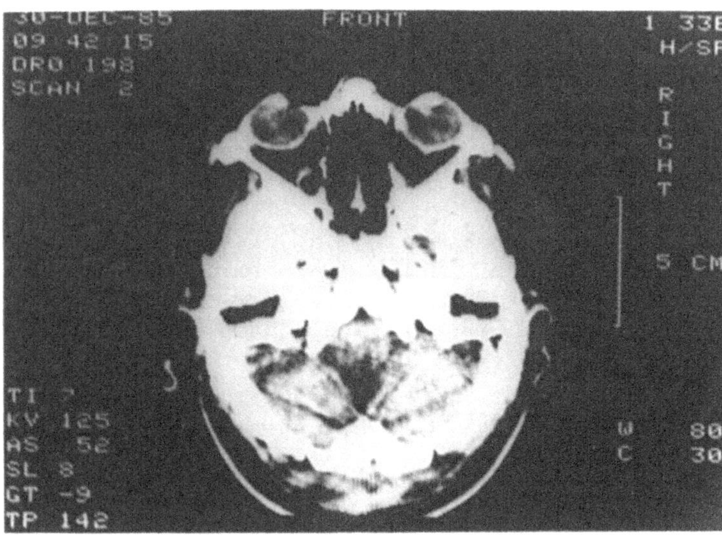

Fig. 14. CT scan 5 months after surgery shows no sign of tumour recurrence

the patient was released home and was able resume his former employment as an engineer and normal family life. The only remaining neurological deficit was paresis of the left hypoglossal nerve. CT control shows no sign of tumour recurrence (Fig. 14).

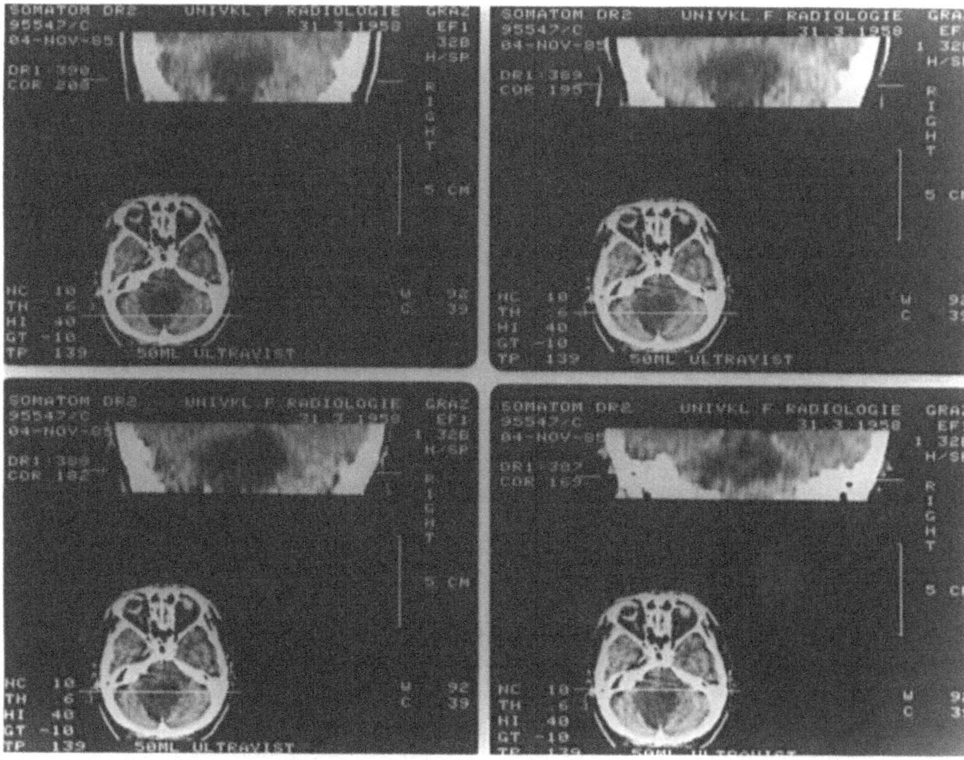

Fig. 15. CT scans show spherical hypodense lesion of fourth ventricle

Case 4: S.H., a 28-year-old male, had a history of occipital trauma (fall) 1 year prior to presentation. Five months later he developed increasing difficulty speaking, vertigo attacks and cerebellar ataxia. A CT scan at another hospital showed a 5 cm × 4 cm spherical tumour in the posterior fossa. The patient was transferred to our clinic with a diagnosis of tumour in the fourth ventricle (Figs. 15 and 16).

Operation: After endotracheal intubation and general anesthesia and with the patient in a sitting position, an osteoplastic craniotomy was performed. The dura mater and arachnoidea were split. In the subarachnoid space we found white, cheese-like masses stemming from a tumour on the lower vermis which reached the cerebellar surface. After splitting of the lower vermis with the CO_2 laser (10 w, focused, cw), the tumour was seen to be well demarcated, reaching into the upper vermis and displacing both cerebellar hemispheres. Caudal dissection was difficult because the cerebellar vasculature was tightly fixed to the tumour and the tumour was inseparable from the floor of the fourth ventricle. Debulking of the tumour was carried out with a Cavitron Ultrasonic Aspirator (CUSA). The affixed

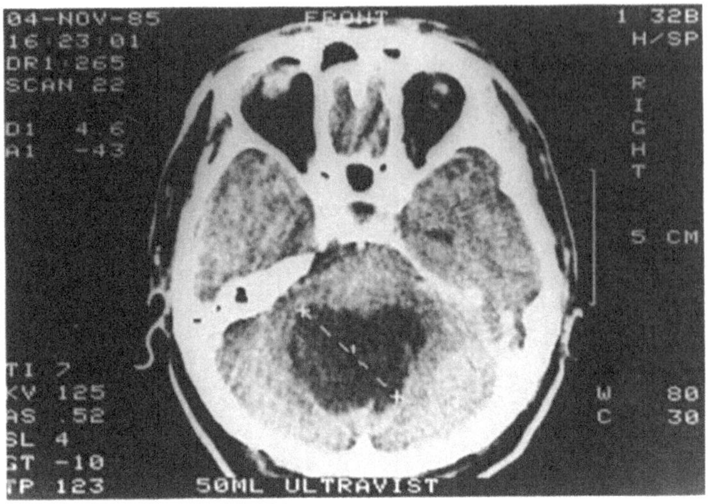

Fig. 16. Closer look at same lesion

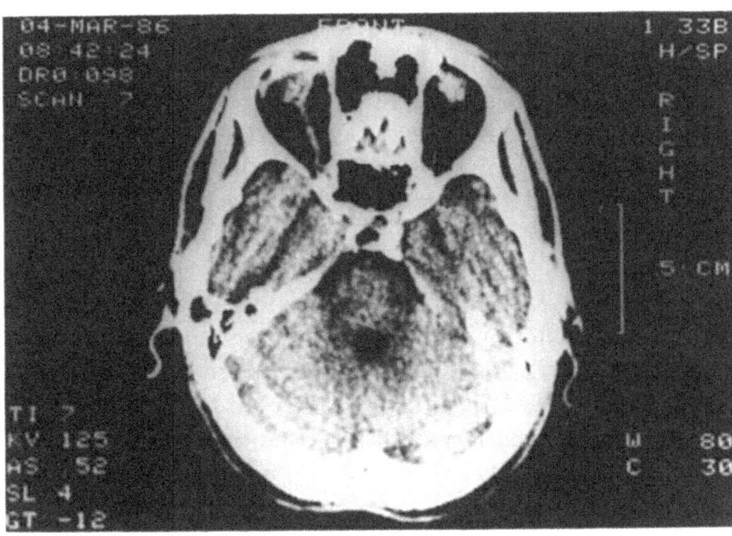

Fig. 17. Postoperative CT scan shows no sign of tumour or recurrence

tumour capsule was vaporized with the CO_2 laser (10 w, medium defocused, cw) as described above. Histology showed an epidermoid cyst. The procedure was completed in the usual fashion; vital functions were stable throughout.

The sutures were removed on the seventh postoperative day. Neurological deficits (cerebellar ataxia) delayed mobilization; difficulty speaking and swallowing increased. Four weeks after surgery the patient was transferred to a rehabilitation center. Five months postoperatively, the patient is about to be released home and a resumption of former employment (baker) is planned. CT control shows no sign of recurrence (Fig. 17).

Conclusion

The introduction of the CO_2 laser significantly expanded the scope of surgery for tumours of the brain stem. Improved diagnostic capabilities (CT, NMR), new operative techniques, the operating microscope, various laser systems and devices such as the CUSA allow us to undertake procedures with the intent to cure which would have been considered impossible 10 years ago.

Today surgeons are using low-watt lasers to weld tissue [108] and fibre-optic laser systems endoscopically [74].

At the Universitätsklinik für Neurochirurgie in Graz, we are showing the feasibility of intra-arterial vaporization of plaques occluding carotid arteries [104]. With a new laser-fibre, we can vaporize the nucleus of a degenerated intravertebral disc after needle puncture under local anesthesia.

Though we have realized our dreams of ten years ago, we are still at the beginning of the laser era. There are still tumours in the brain stem not accessible to surgery—the diffuse gliomas. We are looking for tumour dyes with characteristic absorption spectra in order to then design the corresponding lasers.

References

1. Alvisi C, Cerisoli M, Maccheroni ME (1985) Long term results of surgically treated brainstem gliomas. Acta Neurochir (Wien) 76: 12–17
2. Ascher PW, Oberbauer RW (1977) Laserbeam a new microsurgical instrument. In: Abstract-book (Europ Surg Res, Warsaw), p 66. Karger, Basel, p 66
3. Ascher PW (1977) Longitudinal medial myelotomy with laser. In: Abstract-book (6th Int Congr of Neurol Surgery, Sao Paulo). Excerpta Medica, Amsterdam Oxford, p 110
4. Ascher PW, Oberbauer RW, Holzer P, Knoetgen I (1977) Vorteile und Möglichkeiten des CO_2-Lasers in der Neurochirurgie. Wien Med Wschr 127: 260–262
5. Ascher PW, Oberbauer RW, Clarici G (1977) Laserstrahl — ein modernes neurochirurgisches Instrument. In: Lanner G, Argyropoulos G (eds) Neurochirurgie von heute. Kwizda, Wien, pp 53–57
6. Ascher PW (1977) Laser beam in combination with the micromanipulator: a new aid in microneurosurgery. In: Abstract-book (6th Int Congr of Neurol Surgery, Sao Paulo). Excerpta Medica, Amsterdam Oxford, p 248
7. Ascher PW, Oberbauer RW (1977) Gebrauch des CO_2-Lasers in der Neurochirurgie. In: Abstract-book (Laser 77, Opto-Electronics, München), IPC, Sciences & Techn Press Inc, Richmond, pp 340–342

8. Ascher PW (1977) Longitudinal medial myelotomy with the laser. In: Abstract-book (6th Int Congr of Neurol Surgery, Sao Paulo). Excerpta Medica, Amsterdam Oxford, pp 267–270
9. Ascher PW (1977) Der CO_2-Laser in der Neurochirurgie. Molden, Wien
10. Ascher PW (1978) CO_2-laser, a new neurosurgical instrument. In: Abstract-book (XXI. Int Biennal Congr of the Int College of Surgeons, Jerusalem). p 50
11. Ascher PW (1978) Die Entwicklung des Sharplan 791 CO_2-Lasers zum neurochirurgischen Instrument. In: Abstract-book (19. Tagg Österr Ges f Chir, Wien). H. Egermann, Wien, pp 471–473
12. Ascher PW, Oberbauer RW, Ingolitsch E, Walter GF (1978) Neuere histologische Untersuchungsergebnisse nach Gebrauch des Lasers am Zentralnervensystem. In: Abstract-book (19. Tagg Österr Ges f Chir, Wien). H. Egermann, Wien, pp 479–483
13. Ascher PW (1978) The use of the CO_2-laser in neurosurgery. In: Abstract-book (Laser Surgery, Vol II, Jerusalem). Jerusalem Acad Press, Jerusalem, pp 76–78
14. Ascher PW, Ingolitsch E, Walter GF, Oberbauer RW (1978) Ultrastructural findings in CNS tissue with CO_2-laser. In: Abstract-book (Laser Surgery, Vol II, Jerusalem). Jerusalem Acad Press, Jerusalem, pp 81–90
15. Ascher PW, Heppner F (1978) CO_2-laser a new neurosurgical instrument. Seara Med Neurochirurgica 7: 87–109
16. Ascher PW (1979) Microsurgical use of the CO_2-laser in neurosurgery. In: Abstract-book (Int Med Laser Symp, Detroit). Detroit, p 11a
17. Ascher PW: Laseranwendungen in der Neurochirurgie. In: Abstract-book (Laser 79, Opto-Electronics, München). IPC, Sciences & Techn Press Inc, Richmond, p 5
18. Ascher PW (1979) Laseranwendungen in der Neurochirurgie. In: Abstract-book (Laser 79, Opto-Electronics, München). IPC, Sciences & Techn Press Inc, Richmond, pp 167–169
19. Ascher PW, Heppner F (1979) CO_2-laser surgery of the spinal cord. In: Abstract-book (Laser Surgery, Vol III, Jerusalem). Jerusalem Acad Press, Jerusalem, pp 57–59
20. Ascher PW, Heppner F (1979) Micro laser surgery of the spinal cord. In: Abstract-book (Laser Surgery, Vol III, Jerusalem). Jerusalem Acad Press, Jerusalem, pp 60–61
21. Ascher PW (1979) A neurosurgical CO_2-microlaser. In: Abstract-book (Digest XII Int Conf on Medical & biol Eng and V. Int Conf on Med Physics, Jerusalem) The Combined Meeting Exec Committee, Jerusalem, pp 3–4
22. Ascher PW (1979) A neurosurgical CO_2-Microlaser. In: Abstract-book (Laser Surgery, Vol. III, Jerusalem), Jerusalem Acad Press, Jerusalem, pp 115–118
23. Ascher PW (1979) Last technical developments for the neurosurgical CO_2-Laser. In: Abstract-book (5th Asian-Australian Congr of Neurol Surg, Manila)
24. Ascher PW, Ingolitsch E, Walter GF (1979) Ultrastructural findings of damage and healing of CNS tissue after use of CO_2-laser in comparison to other

techniques. In: Abstract-book (5th Asian-Australian Congr of Neurol Surg, Manila)
25. Ascher PW, Sager WD (1979) Computertomographie-Kontrollen nach Laser-Operationen. In: Computertomographie. Thieme, Stuttgart, pp 10–14
26. Ascher PW, Heppner F (1979) Laser in medicine and neurosurgery. In: Abstract-book (Progetto Finalizzato Laser di Potenza, Florenz), pp 22–28
27. Ascher PW (1979) Advantages and limitations of the CO_2-Laser in neurosurgery. In: Abstract-book (Int Conf on Lasers 79, Techn Digest, Orlando), p 11
28. Ascher PW (1979) Advantages and limitations of the CO_2-Laser in neurosurgery. In: Abstract-book (Int Conf on Lasers 79, Orlando), STS-Press, McLean, pp 122–125
29. Ascher PW, Heppner F, Oberbauer RW, Walter GF, Ingolitsch E (1979) Laserbeam—a new microsurgical instrument. In: Abstract-book (Int Conf on Lasers 79, Orlando). STS-Press, McLean, pp 696–699
30. Ascher PW (1980) Der CO_2-Laser in der Neurochirurgie. Fortschr d Med 98: 253–254
31. Ascher PW, Holzer P (1980) Chirurgie der peripheren Nerven mit dem CO_2-Laser, verglichen mit herkömmlichen Instrumenten. Zbl Neurochir 41: 37–42
32. Ascher PW (1980) The value of recent advances using the CO_2-laser in neurosurgery. In: Abstract-book (Int Conf on Lasers, Conf Guide, Shanghai), p 20
33. Ascher PW (1980) The value of recent advances using the CO_2-laser in neurosurgery. In: Abstract-book (Int Conf on Lasers, Digest of Techn Papers, Shanghai), p 148
34. Ascher PW, Heppner F (1980) Clinical applications of lasers in neurosurgery. In: Lasers in medicine. John Wiley & Sons, Chichester, New York, pp 1–14
35. Ascher PW (1980) The value of recent advances in radiographic and neurosurgical techniques in the removal of spinal canal lesions. In: Radiographic evaluations of the spine. Masson Publ Inc, New York, pp 717–721
36. Ascher PW (1982) Chapter 23: Neurosurgery. In: Microscopic and endoscopic surgery with the CO_2-laser. John Wright PSG Inc, Boston, pp 298–314
37. Ascher PW (1981) CO_2-laser: An instrument for surgery beyond the borderline of operability. In: Techn Digest (Cleos, 1981, San Diego), p 148
38. Ascher PW (1981) Removal of intramedullary tumors by laser vaporization. In: Abstract-book (7th Int Congr of Neurol Surgery, München). Thieme, München, p 160
39. Ascher PW, Heppner F (1982) The CO_2-laser in neurosurgery. In: Abstract-book (Int Adv in Surg Oncology, Vol V, New York). Alan R. Liss Inc, New York, pp 385–396
40. Ascher PW (1981) Lasers in neurosurgery, history and development. In: Abstract-book (Lasers' 81, Tokyo), Inter Group Corp, Tokyo, p 4, p 20
41. Ascher PW (1981) Early studies in neurolaser interaction with Nd:YAG. In: Abstract-book (1st Am Congr on Laser Neurosurg, Chicago)

42. Ascher PW (1981) Micro laser surgery of the spinal cord. In: Abstract-book (1st Am Congr on Laser Neurosurg, Chicago), p 54
43. Ascher PW (1982) Micro laser surgery of the spinal cord. In: Abstract-book (Scand Neurosurg Soc, 34th Ann Meeting, Trondheim), p 98
44. Ascher PW, Oberbauer RW, Heppner F (1982) Surgery of tumors in the midline. In: Abstract-book (8th Congr Europ Soc for Paediatric Neurosurgery, Rennes), p 24
45. Ascher PW (1982) Different lasers in the therapy of gliomas. In: Abstract-book (Int Symp Advances in Oncology: Diagnosis and Therapy on Bone & Brain Tumors, Sorrento)
46. Ascher PW (1983) Absolute indications for laser use in neurosurgery. In: New frontiers in laser medicine and surgery. Excerpta Medica, Amsterdam Oxford, pp 181–187
47. Ascher PW (1983) Absolute Indikationen für den CO_2- und Neodymium-YAG-Laser-Einsatz in der Neurochirurgie. Fortschr Med 101: 1033–1036
48. Ascher PW, Cerullo L (1983) Chapter 9: Laser use in neurosurgery. In: Surgical application of lasers, Year Book. Med Publ, Chicago, pp 163–174
49. Ascher PW (1983) Biological and histological findings using different lasers. In: Abstract-book (I. Congresso Panamericano de Cirurgia con Laser, Cartagena)
50. Ascher PW, Marguth F, Beck OJ, Heppner F, Frank F (1983) The Nd: YAG-laser in neurosurgery. In: Abstract-book (I. Congresso Panamericano de Cirurgia con Laser, Cartagena), p 21
51. Ascher PW (1983) Surgery of brain stem tumours in children. In: Abstract-book (I. Congresso Panamericano de Cirurgia con Laser, Cartagena), p 28
52. Ascher PW (1983) Micro laser surgery of the spinal cord. In: Abstract-book (I. Congresso Panamericano de Cirurgia con Laser, Cartagena), p 31
53. Ascher PW (1983) Status quo and new horizons of lasers in neurosurgery. In: Abstract-book (VIII. Congresso Mexicano de Cirurgia Neurologica, Acapulco), p 174
54. Ascher PW (1983) Different lasers in neurosurgery: a comparison of Neodymium-YAG- and CO_2-laser. In: Neodymium: YAG-laser in medicine and surgery. Elsevier, New York, pp 119–125
55. Ascher PW (1983) Report on the use of the Cusa system in neurosurgery. Surg Update 1: 3, 2–4
56. Ascher PW (1983) Recent advances of various lasers in neurosurgery. In: Abstract-book (7th Europ Congr of Neurosurgery, Brussels)
57. Ascher PW, Clarici G, Auer L, Walter G (1983) Functional hypophysectomy with lasers. In: Abstract-book (7th Europ Congr on Neurosurgery, Brussels), p 239
58. Ascher PW, Clarici G (1983) Different lasers used for transsphenoidal hypophysectomy. In: Abstract-book (7th Europ Congr on Neurosurgery, Brussels), pp 163–168
59. Ascher PW (1983) Status quo of laser use in neurosurgery. Neurologia Colombiana 7: 163–168
60. Ascher PW (1983) Recent advances in the use of the CO_2-laser in neurosurgery. In: Abstract-book (Int. Conf on Laser, Beijing-Shanghai), Schina Acad Publ & John Wiley & Sons Publ, Beijing Boston, pp 737–742

61. Ascher PW (1984) Lasers in neurosurgery. In: Abstract-book (2nd Int Laser Symp Bern)
62. Ascher PW (1984) Lasers in neurosurgery. In: Abstract-book (2nd Int Laser Symp Bern), Laser Med & Surg News, Vol II, 1, Mary Ann Lieberth Inc Publ, New York, p 10
63. Ascher PW (1984) Seldom done operations for pain relief using various lasers. In: Abstract-book (Int Symp on Advances in Pain Research & Therapy, Lakenhof), p T-II/7
64. Ascher PW (1984) Transdural denaturation of pituitary tissue with the Nd:YAG-laser. In: Abstract-book (3rd Congr on Laser Neurosurgery, Chicago)
65. Ascher PW (1984) Absolute inoperable tumors has to be reconsidered under the use of different lasers. In: Abstract-book (3rd Congr on Laser Neurosurgery, Chicago), p 192
66. Ascher PW, Heppner F (1984) CO_2-laser in neurosurgery. Neurosurg Rev 7
67. Ascher PW (1984) Verschiedene Laser als neurochirurgische Instrumente. In: Medizin und Technik, Styria, Graz, pp 129–131
68. Ascher PW, Oberbauer RW (1984) Laser, microscope and endoscope in central brain lesions. Neuropediatrics 15: 234
69. Ascher PW (1984) Lasers in neurosurgery—status of art. In: Abstract-book (2nd Seminar in Laser and Medical Appl, Taipeh)
70. Ascher PW (1984) Laser in neurosurgery, present and future. In: Handbook Laser Workshops, Singapore, p 77
71. Ascher PW (1984) Laser in der Neurochirurgie. Soziale Berufe 36: 8–11
72. Ascher PW, Clarici G (1981) Funktionelle Hypophysektomie mit dem Laser. In: Verhandlungsberichte d Dt Ges f Lasermed e V, München. Erdmann Brenger, München, pp 126–129
73. Ascher PW (1985) Anwendung verschiedener Laser in der Neurochirurgie. In: Verhandlungsberichte d Dt Ges f Lasermed e V, München. Erdmann Brenger, München, pp 106–114
74. Auer LM, Ascher PW (1983) Endoscopic burrhole evacuation of intracerebral hematomas using ultrasound morcellement and microlaser coagulation. In: Abstract-book (7th Europ Congr Neurosurg, Brussels), p 129
75. Cole FM, Yates PO (1967) The occurrence and significance of intracerebral micro-aneurysms. J Path Bact 93: 393–411
76. Dandy WE (1945) Surgery of the brain. In: Lewis (ed) Practice of surgery, Vol XII, W F Prior Co, Inc, Magerstown
77. Entzian W (1984) Removal of low brain stem glioblastomas—positive long term results in circumscribed lesions. In: Abstract-book (12th Scient Meeting Int Soc Pediatr Neurosurgery, Cairo)
78. Heppner F (1978) Das Laserskalpell am Nervensystem. Wien Med Wschr 128: 7, 197–201
79. Heppner F (1978) Neurochirurgische Operationen mit dem CO_2-Laser. Neurologia et Psychiatria 1: 1, 9–13
80. Heppner F (1979) Behandlung von Mittellinien-Laesionen des Gehirns mit

dem CO_2-Laser. In: Abstract-book (19. Tagg Öst Ges f Chir, Wien), Egermann, Wien, pp 446–447

81. Heppner F (1978) The laser scalpel on the nervous system. In: Kaplan (ed) Laser surgery II, Jerusalem Acad Press, Jerusalem, p 79
82. Heppner F (1979) Erfahrungen mit dem CO_2-Laser in der Chirurgie des Nervensystems. Zentralbl Neurochir 40: 297–304
83. Heppner F (1980) Neuere Technologien im Einsatz gegen hirneigene bösartige Geschwülste. Proceedings. Österr Akad Wiss, Wien
84. Heppner F (1981) Der Laser in der Neurochirurgie. In: Dinstl, Fischer (eds) Der Laser. Springer, Berlin Heidelberg New York, p 143
85. Heppner F (1983) Lasers in neurosurgery and their forerunners. In: Atsumi (ed) New frontiers in laser surgery & med. Excerpta Medica, Amsterdam Oxford Princeton, p 39
86. Heppner F (1982) Der medizinische Laser und seine Vorläufer. In: Staehler, Hofstetter (eds) Verhandlungsbericht d Dt Ges f Lasermed e V, München. W. Zuckschwerdt, München, pp 1–8
87. Heppner F (1983) Die heilende Hitze. Fortschr Med 101: 14, 621
88. Heppner F (1984) Die medizinischen Laser — heute. Öst Ärzteztg 39: 11, 811
89. Heppner F (1983) Further advances in laser surgery of the central nervous system. Lasers Surg Med 3: 2, 108
90. Heppner F (1984) The glioblastoma multiforme—a lifelong challenge to the neurosurgeon. In: Abstract-book (Scand Neurosurg Soc, Stockholm)
91. Heppner F (1984) Das Machbare und der Laser. Fortschr Med 23: 629
92. Heppner F (1984) CO_2-laser in neurosurgery. Neurosurg Rev 7: 123
93. Heppner F (1984) Direct surgical attack to pontine and rhombencephalic lesions. In: Abstract-book (Europ Soc for Paediatr Neurosurg, Wien)
94. Heppner F, Schuy St, Ascher PW, Holzer P, Wiesspeiner G (1985) Lasers and telethermy in the treatment of glioblastoma multiforme. In: Abstract-book (Topics in Brain Tumours, Verona)
95. Heppner F, Ascher PW (1976) Über den Einsatz des Laserstrahls in der Neurochirurgie. Medizinalmarkt 12: 424
96. Heppner F, Ascher PW (1977) Hirnoperationen mit dem CO_2-Laser. Mels Med Mitt [Suppl] II: 51, 121
97. Heppner F, Ascher PW (1977) Erste Versuche mit dem Laserstrahl in der Behandlung neurochirurgischer Erkrankungen. Zentralbl Neurochir 38: 77
98. Heppner F, Ascher PW (1977) The use of laser in neurosurgery. In: Abstract-book (6th Int Congr of Neurol Surg, Sao Paulo), p 134
99. Heppner F, Ascher PW (1977) Operationen an Hirn und Rückenmark mit dem CO_2-Laser. Acta chir Austr 9: 32–34
100. Heppner F, Ascher PW (1984) Laseroperationen. In: Schmitt (ed) Tumoren der Wirbelsäule. Hippokrates, Stuttgart, p 161
101. Heppner F, Oberbauer RW, Ascher PW (1985) Direct surgical attack on pontine and rhombencephalic lesions. Acta Neurochir (Wien) [Suppl] 34: 123–125
102. Hoffman HJ, Becker L, Craven MA (1980) A clinical and pathologically distinct group of benign brain stem gliomas. Neurosurgery 7: 243–248

103. Koos WT, Miller MH (1971) Intracranial tumours of infants and children. Thieme, Stuttgart
104. Lammer J, Ascher PW, Choy DSJ (1986) Transfemorale Katheter-Laser-Thrombendarteriektomie der Arteria carotis. DMW 16: 607–610
105. Lassman LP, Arjona VE (1967) Pontine gliomas of childhood. Lancet 1: 913–915
106. Lassman LP (1974) Tumours of the pons and medulla oblongata. In: Vinken, Bruyn (eds) Handbook of clinical neurology, Vol XVII, Chapter 19. North Holland Publ Comp, Amsterdam
107. Matson DD (1969) Neurosurgery of infancy and childhood, 2nd ed. Ch C Thomas, Springfield, Ill
108. Neblett CR (1985) Laser vascular application: reconstructive and reparative. In: Fasano (ed) Advanced intraoperative technologies in neurosurgery. Springer, Wien New York
109. Oberbauer RW, Heppner F, Ascher PW (1982) Surgery of tumours in the midline. In: Abstract-book (8th Meeting Europ Soc Pediatr Neurosurgery, Rennes), p 24
110. Obrador S, Dierssen G, Odoriz BJ (1970) Surgical evacuation of a pontine medullary haematoma. Case Report. J Neurosurg 33: 82–84
111. Panitsch HS, Berg BO (1970) Brain stem tumours of childhood and adolescence. Am J Dis Child 119: 465–472
112. Russel DS, Rubinstein LJ (1971) Pathology of tumours of the nervous system. Edward Arnold Ltd, London
113. Strange P, Wohlert L (1982) Primary brain stem tumours. Acta Neurochir (Wien) 62: 219–232
114. Walker EA, Hopple TL (1949) Brain tumours in children. J Pediat 35: 671–687
115. Walker M (1981, 1983, 1984, 1985) Personal informations. Salt Lake City

Address for correspondence: Prof. Dr. Peter Wolf Ascher, Universitätsklinik für Neurochirurgie Graz, A-8036 Graz, Austria.

Reconstructive Vascular Neurosurgery: Microsurgical CO Laser Application

C. R. Neblett

Houston, Texas, U.S.A.

Introduction

No one relishes a stringent challenge more than a neurological surgeon. And no arena fulfills that criteria more completely than intracranial vascular reconstruction.

In the continuing quest to establish or maintain critical blood flow, our armamentarium of tools must be expanded. Applications range from microscopes to microsutures, from microinstruments to microsurgical lasers.

The neurological surgeon must contend with small calibre vessels, arteries of diminished structural integrity, lack of collateral blood flow, and frequently compromised accessibility.

This chapter considers the question of applicability of the CO_2 laser used in the micro mode for the reconstruction of vascular tissue.

The Project

A feasibility project was developed. The initial question was: Can organic tissue be bonded through exposure to the thermal effects of the CO_2 laser? Mr. James Morris*, who developed the Bio Quantum Technologies microsurgical CO_2 laser Model No. 7600, and I selected the most difficult structure for reconstruction we could imagine, a small artery. An artery of approximately one millimeter has been considered the minimum diameter for human clinical application. The experimental model, the femoral artery of the rat, allowed immediate results to be obtained. Patency and degree of blood leak-free status were immediately observable. Follow-up studies provided an assessment of healing qualities: Did the arterial edges remain in proximity? Was patency maintained? What was the integrity level? What was the histological quality of healing?

* Mr. James Morris, Bio Quantum Technologies, Inc., Houston, Texas.

The limitations of this study were many. The variables were multiple. The conclusions were preliminary.

However, the basic question was, "Can organic tissue be bonded through exposure to the thermal effects of microsurgical CO_2 laser technique?" The answer: "Yes."

Once this was established, the analysis of the limitations and variables was begun. Included among these were: the type of experimental animal, the type of artery (elastic/muscular), the size of the artery, the end-to-end anastomosis, the end-to-side anastomosis, surgical technician variability, the laser instrument variability, the type of laser, the mode of application, the wattage, the spot size, the time exposures, the methods of post surgical evaluation, etc. These topics will be reviewed.

Experimental Animal Model: The classic animal model, the rat, was chosen. Significant literature data was available for comparison. Their expense and care were reasonable. Other animals included in this study were the rabbit, the dog, the cat, the pig and the monkey.

Artery Type: The femoral artery was used for the base study. The carotid, the aorta and the veins were also bonded. Both predominantly elastin and predominantly muscularis artery types were fused. However, the thermal effect on each type of cell is variable and must be evaluated.

Arterial Size: Small arteries (0.5 to 1.5 mm) were desired. Up to 3 mm arteries have been successfully microsurgically CO_2 laser anastomosed.

End-to-End, End-to-Side: Both procedures were performed successfully. End-to-end was the standard.

Surgical Technicians: Multiple surgeons performed these anastomoses. Results were somewhat variable dependent upon: the individual, individual skill, microsurgical experience, CO_2 application experience, and personal preference of CO_2 laser parameters applied.

Laser Type: The CO_2 laser (10.6 microns) was the only wave length evaluated in this study. The only instrument used was the Bio Quantum Technologies microsurgical CO_2 laser Model No. 7600.

Application Modes: Delivery was in the micro mode. Precision required microscopic magnification. Accurate application and ability to easily visualize tissue changes during fusion were essential.

Wattage: Fifty to 150 milliwatts usually were satisfactory for the femoral artery of the rat. Variations were dependent upon the type of artery and its diameter. A rule of thumb: take the diameter of the artery in millimeters and multiply it by 100 to determine the approximate milliwatt level.

Spot Size: The spot size was important. One hundred fifty microns at a working distance of 300 mm was standard for the rat femoral artery. Microns ranging from 80 to 500 mm were applied and varied with the working distance as well as the character and diameter of the vessel. Certainly its variation altered power density. The spot size determined the amount of tissue proximal and distal to the exact anastomotic site which was exposed to thermal factors. Also the surgical application of this spot size influenced the area of artery treated. These factors affected the healing process.

Time Exposure: The time exposures were variable. The technique most frequently applied was at 0.1 to 0.3 second intervals. I personally preferred the continuous mode.

Post Surgical Studies: The histologic studies were done under the direction of Sharon Thomsen, M.D.* Those included standard light microscopic and scanning electron microscopy. Tensile strength evaluations were performed. The histological, mechanical and biochemical changes continue to require intense analyzation.

Technique

The Sprague-Dolly rat with an average weight of 300 grams was utilized. Intraperitoneal Ketamine—Hcl NF was applied for anesthesia. The femoral artery was exposed through a groin incision as frequently described in the literature. It was exposed between the inguinal ligament and the superficial epigastric vessel and the profundus femoralis branch cauterized and cut with the laser. An Acland clamp was applied and the artery transversely severed. The open lumen was lavaged with physiologic saline to remove all blood particles.

Three stay stitches of 10/0 Dermalon were placed 120° apart to provide approximation of the arteries' ends. The sutures were left long enough to be easily grasped with forceps, thus serving as stay stitches. Light outward tension applied to two of these stitches provided good approximation of the proximal and distal ends of the artery. Firm approximation was essential.

Working at a distance of 300 millimeters through the surgical microscope at 16 to 25 power, the approximated edges were exposed to the CO_2 laser. The spot size was 150 microns and the milliwatts were approximately 100, depending upon the size and other physical characteristics of the vessel. Used in a continuous mode, or an intermittent mode (0.1 to 0.3 seconds), the tissue was fused. The vessel would then be rotated and each third of the vessel done consecutively. Physiologic saline was applied to the vessel regularly. Accomplished fusion was visually observed. The anastomotic site appeared slightly "darkened" and "dried". The arterial ends were securely together. Constriction of the vessel or a significant darkening of tissue is an indication of excessive exposures.

The distal clamp was removed first and back flow observed. If a leakage of blood was noted the clamp was reapplied. This blood was lavaged from

* Associate Professor, Department of Pathology, University of Miami, Miami.

the lumen and from the interface between the arterial walls. After tension was applied to the stay stitches for approximation purposes, a re-exposure to the laser with the above-described parameters was introduced. The clamps were again removed, first distal and then proximal. A standard patency test was performed. The incisional site was closed in a standard manner.

Results

Patency rates varied from 90% to 100%. Aneurysmal formation at the anastomotic site was found in 0% to 13% of the specimens. As noted in the previous pages, these variables reflected the results from different surgeons, their experience, and the application of different techniques.

In reviewing our results, some general conclusions can be made. In arteries of approximately 1 mm diameter, a long-term patency rate of greater than 95% can be expected.

Aneurysmal formation did occur. Perhaps this was one of the better mechanisms for establishing precision of technical application. Firm approximation of the arterial ends was critical. This can be accomplished either using three or four approximating stitches. James I. Ausman, M.D.* has demonstrated the benefit of four stitches. Stripping of the adventitia should be avoided. The CO_2 laser energy should be applied directly to the anastomotic site. Either too large a spot size or too great an excursion of the laser to either side of the anastomosis can result in undesirable thermal exposure resulting in chemical and structural changes. After completion of the bonding, if blood leakage occurred, it was essential that all blood particles be lavaged from that opening prior to reapplication of the laser. Retained blood resulted in a "blood bond" which did not provide adequate structural integrity. Adherence to strict technical practices should significantly minimize this aneurysmal formation.

The tensile strength of the anastomosis was also evaluated. The Intron Universal Materials Testing Instrument, Model 1125, with a 100 gram load cell was used to measure the tensile strength. Evulsion of the artery occurred at a mean value of: 19.5 grams—immediately postoperative; 20.6 grams—third postoperative day; 34.6 grams—seventh postoperative day; and 95 grams—twenty-first postoperative day. A normal rat femoral artery ruptures at a mean of 95.3 grams. This study indicated adequate strength initially and almost normal tensile strength by the third postoperative week.

Histology

The histological assessment of the laser assisted microvascular anastomosis was evaluated by Dr. Sharon Thomsen. Both light microscopy and scanning electron microscopy were utilized.

---

* Department of Neurosurgery & Neurology, Henry Ford Hospital, Detroit, Michigan.

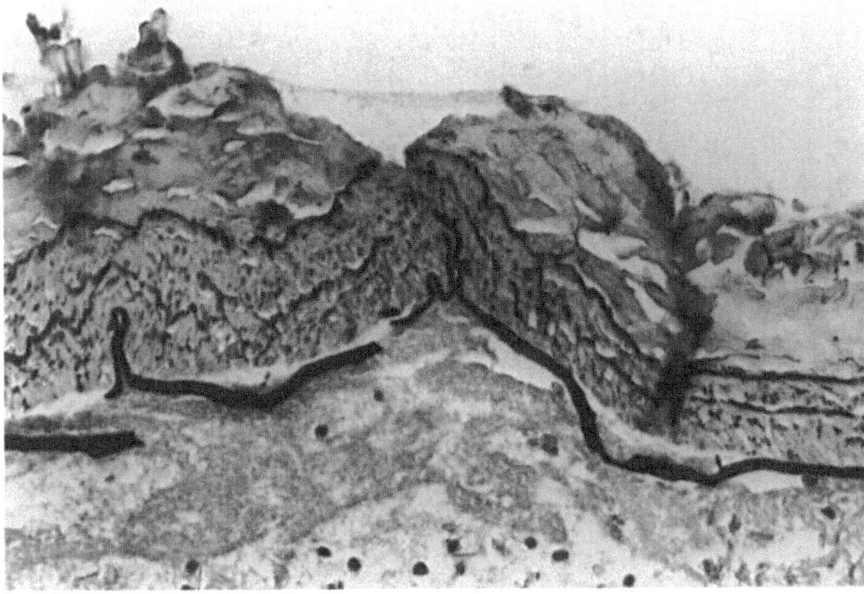

Fig. 1. The initial primary anastomic bond appears to be a thermal coagulation of adventitial and medial tissues (immediate)

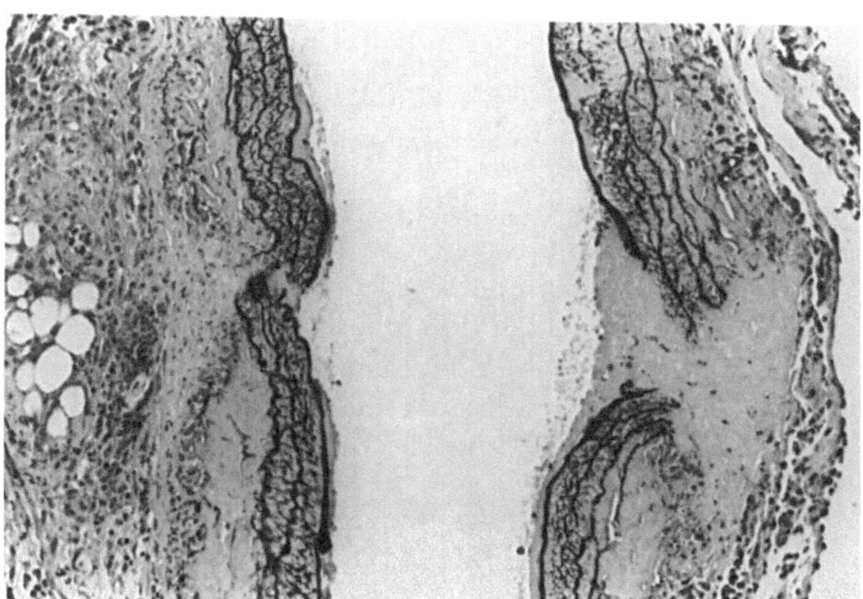

Fig. 2. The wound edges were filled with fibrin which formed a blood-tight plug in the wound (three days)

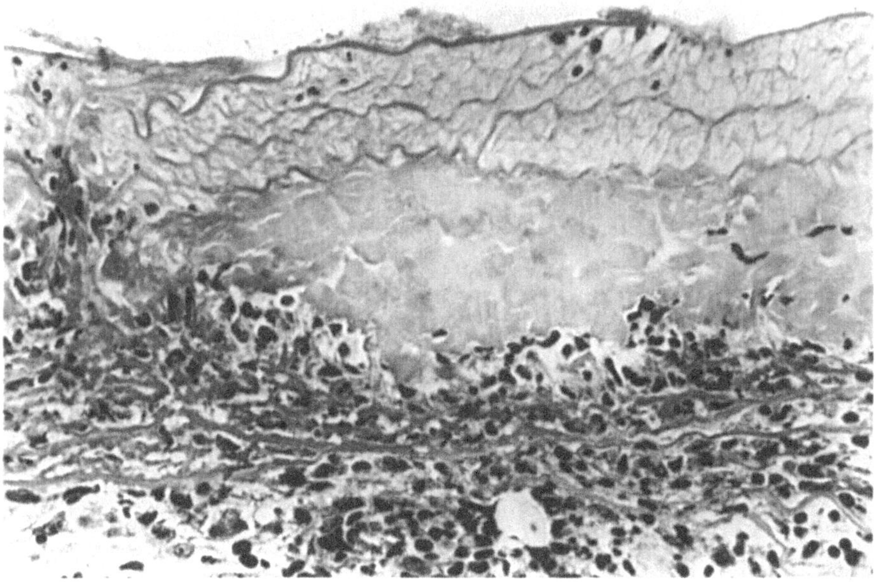

Fig. 3. The adventitial coagulum and the adventitial portion of the fibrin had become organized by an infiltrate of macrophages, lymphocytes and polymorphonuclear cells (three days)

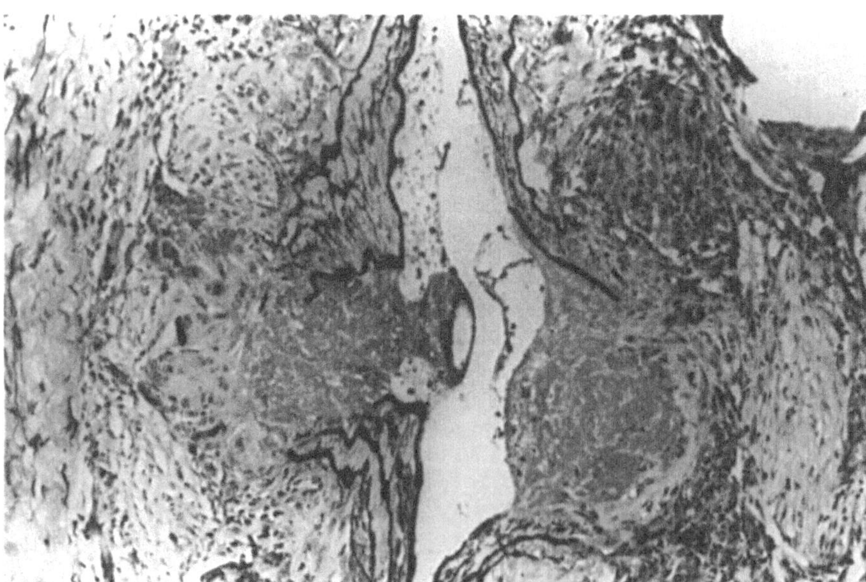

Fig. 4. The formation of vascular granulation wound tissue began in the adventitial aspect of the wound (one to two weeks)

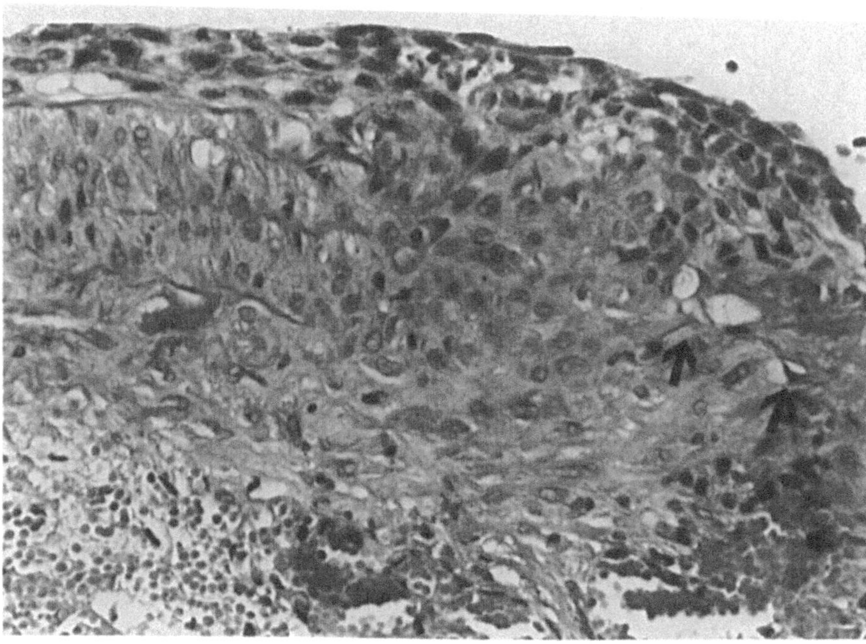

Fig. 5. The re-endotheliozation seemed to be complete and the internal cells lying beneath the endothelium extended down into the small amount of fibrin remaining within the wound gap (three weeks)

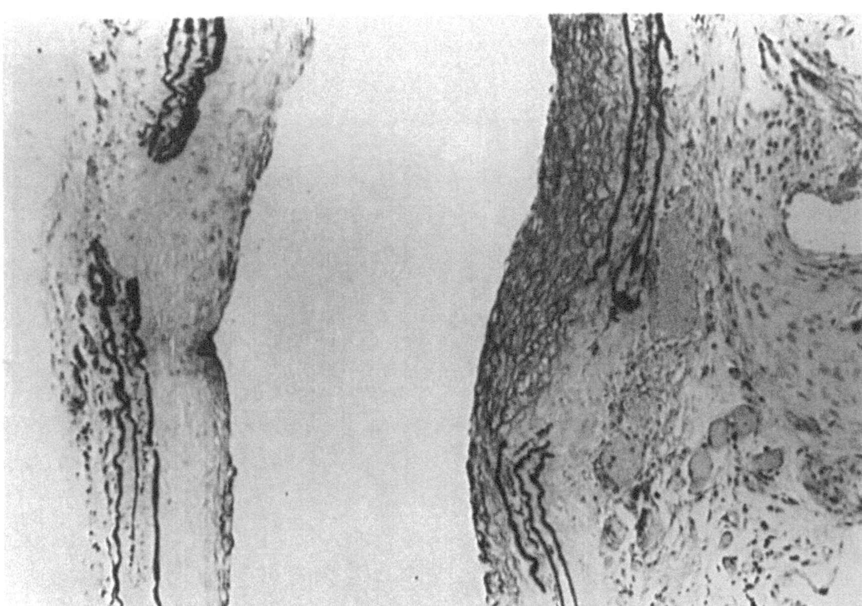

Fig. 6. The wound healing was essentially complete with further organization of the vascular granulation tissue and the intimal tissue and more collagen laid down by the cells in the wound (five to seven weeks)

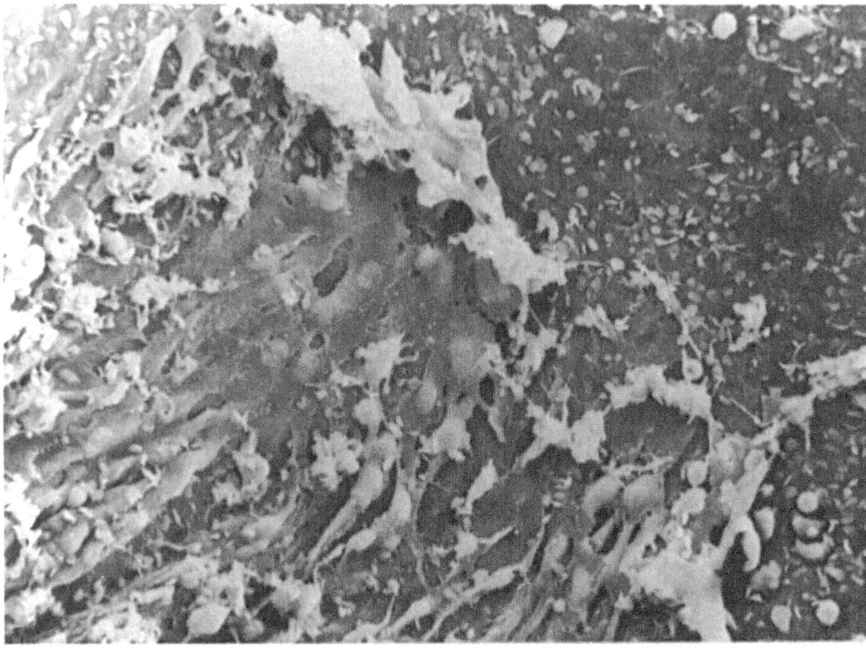

Fig. 7. Scanning electron microscopy demonstrated a thin film of fibrin covering the wound edges and the underlying interelastic membrane exposed and covered with scattered adherent platelets (one hour)

Fig. 8. Re-endotheliozation had begun with proliferation of endothelio cells both at the extreme boundaries of the intact endothelium and among the islands of endothelium closer to the anastomotic site (five days)

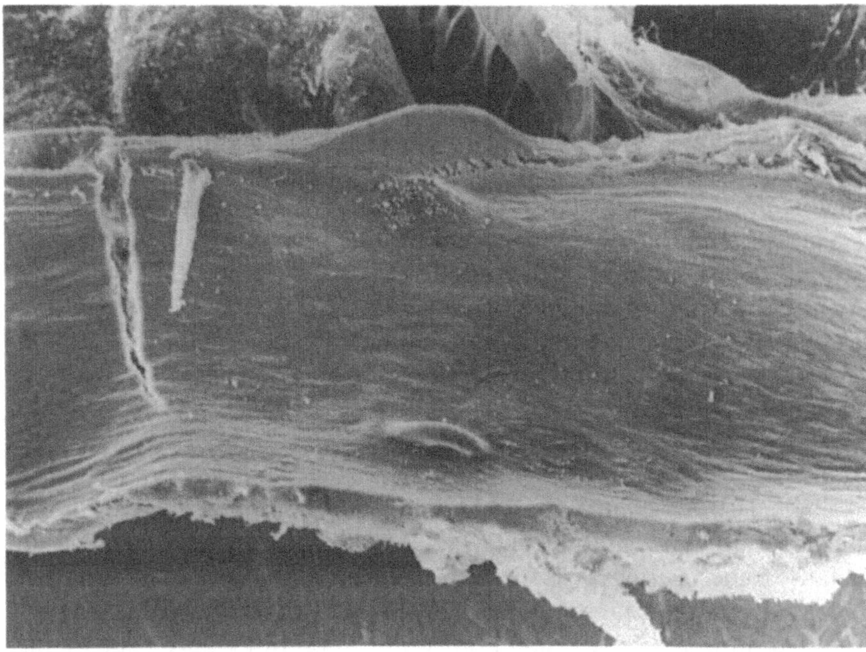

Fig. 9. Re-endotheliozation continued and most anastomoses were completely covered with intact continuous endothelium (three weeks)

Specimens studied immediately after laser assisted microvascular anastomoses demonstrated that the primary anastomotic bond appeared to be a thermal coagulation of adventitial and medial tissues (Fig. 1). Exposure to the laser did not seem to change the internal or external elastic lamina.

At three days the wound edges were filled with fibrin which formed a blood-tight plug in the wound (Fig. 2). The adventitial coagulant and the adventitial portion of the fibrin had become organized by an infiltrate of macrophages, lymphocytes and polymorphonuclear cells (Fig. 3).

At one to two weeks the formation of vascular granulation wound tissue began in the adventitial aspect of thc wound (Fig. 4). The internal portion of the wound gaps still contained fibrin undergoing organization predominantly progressing from the adventitial to the luminal surface.

At three weeks the re-endothelialization seemed to be complete and the intimal cells lying beneath the endothelium extended down into the small amount of fibrin remaining within the wound gap (Fig. 5). Cells which resemble intimal cells were present in the interstices of the medial scaffolding immediately adjacent to the wound edges.

At five to seven weeks the wound healing was essentially complete with further organization of the vascular granulation tissue in the intimal tissue and more collagen laid down by the cells in the wound (Fig. 6).

Scanning electron microscopy demonstrated within the first postoperative hour a thin film of fibrin covered the wound edges. The underlying internal elastic membrane became exposed and was covered with scattered adherent platelets (Fig. 7).

At one day most of the damaged endothelium had sloughed, leaving a peripheral boundary and residual islands of intact endothelium in the vessel.

At five days re-endothelialization had begun with proliferation of endothelial cells both at the extreme boundaries of the intact endothelium and among the islands of endothelium closer to the anastomotic site (Fig. 8).

At three weeks re-endothelialization continued and most anastomoses were completely covered with intact continuous endothelium (Fig. 9).

By four weeks the luminal healing was complete and no further pertinent changes were seen.

Conclusion

Neurological surgery must meet the challenge presented by pathologic problems resultant from intracranial blood flow impairments. The potential of including microsurgical CO_2 laser principles should be a consideration.

Material has been presented to help initiate the evaluation of applicability. Additional studies, definitions and alterations may lead to the blending of high technology with surgical procedures which will result in improved techniques and, therefore, an enhanced quality of life.

Reference

1. Neblett CR, Morris JR, Thomsen S (1986) Laser-assisted microsurgical anastomosis. 19: 914–934

Address for correspondence: C. R. Neblett, M.D., P.A., 1748 Scurlock Tower, 6560 Fannin Street, Houston, TX 77030, U.S.A.

Laser Applications in Pediatric Neurosurgery

K. R. Crone, T. S. Berger, and **J. M. Tew Jr.**

Children's Hospital Medical Center, Department of Pediatric Neurosurgery, and University of Cincinnati Medical Center, Department of Neurosurgery, Cincinnati, Ohio, U.S.A.

Introduction

The application of lasers in neurosurgery represents one of the most significant technological advances in the development of tools for precise micro-dissection. When combined with the operating microscope and micro-neurosurgical principles, the use of a laser provides controlled dissection of the interface between the lesion and normal nervous tissue without excessive manipulation or retraction of adjacent structures. Such operative precision is particularly useful in pediatric neurosurgery where lesions commonly involve the brain stem, fourth ventricle, optic pathways, suprasellar region, or spinal cord.

Increased experience with lasers in pediatric neurosurgery has provided information regarding the indications for its applications [3, 4, 6–8, 11, 12]. In this chapter we will briefly describe laser light principles, the lasers available for neurosurgical use, and finally illustrate clinical situations in pediatric neurosurgery where we have found the laser to be of particular benefit.

Laser Principles

The laser is an acronym for Light Amplification by Stimulated Emission of Radiation. Laser light is produced by stimulation of a lasing medium which may contain either a gas (argon, carbon dioxide) or a solid crystal (ruby, neodymium: YAG). Energized molecules within the medium undergo spontaneous decay emitting photons of a uniform wavelength, thereby, producing a monochromatic source of light energy. Since these photons from a given laser medium are all in sinusoidal phase with one another and travel parallel within the laser beam both temporal and spatial coherence occurs. As a result of these characteristics, light energy of the laser can be focally concentrated through a series of lenses to provide the ability to cut, coagulate, or vaporize tissue. The biological effect on tissue is generated by a thermal reaction derived through excitation of vibrational and rotational levels of tissue matter.

The interaction between laser light and tissue is dependent upon the type of laser (wavelength) and the biological characteristics (cell density, water content, pigmentation) of the tissues to which it is applied. Prediction of a given laser tissue interaction is inprecise due to the variability of the previously mentioned tissue characteristics. Such variability, however, does not preclude generalizations about specific effects of different lasers.

Neurosurgical Lasers

The carbon dioxide, argon, and neodymium:YAG lasers are currently approved for clinical use in neurosurgery. The specific wavelength for each laser affects tissue differently and knowledge of such effects are essential to proper selection of the laser for clinical use.

CO_2 Lasers

The carbon dioxide laser emits a wavelength of 10.6 microns which is located within the far infrared region of the spectrum. Energy of this wavelength is rapidly absorbed by water. Near total surface absorption occurs in neural tissue which is composed of 70–80% water. Little thermal damage is produced at adjacent tissues from a focally concentrated CO_2 laser beam as 98% of the CO_2 laser pulse is absorbed in the first 200 microns of tissue penetration and under most circumstances the zone of injury surrounding the zone of the CO_2 laser beam is less than one millimeter in width [1, 2].

The initial effect of CO_2 on tissue is thermal coagulation, thus, appropriate adjustment of the focal point can effectively coagulate small tumor vessels. Additional energy absorption within the tissue produces vaporization of the coagulate neoplastic tissue.

The power of the CO_2 laser beam can be varied from microwatts to 100 watts. Such a versatile range of power combined with an adjustable spot size and exposure time permits precise tissue dissection, coagulation, and vaporization.

Argon Laser

The wavelength emitted from the argon laser is located in the blue-green visible spectrum at 488–514.5 microns. This wavelength is absorbed by colored or pigmented tissue and readily passes through water or a clear medium. These features would appear suitable for applications of the argon laser to vascular tumors of the nervous system but the small amount of energy generated from this laser and the necessity for sharp focusing to provide energy vaporization of tissue has proved the argon laser to be less useful than the CO_2 laser.

At present, the greatest potential use for the argon laser in neurological surgery appears to be a technology designed for endoscopic fenestration of cysts or intraventricular adhesions. A recent report by Powers involving two patients in which endoscopic techniques were used provide encouraging results for further developing these techniques [9].

Nd: YAG Laser

Until the early part of 1986, the YAG laser required an Investigational Device Exemption from the Food and Drug Administration. The YAG laser emits a wavelength of 1.06 microns which is located in the near infrared spectrum. Similar to the argon laser it is poorly absorbed by non-pigmented tissue, passes well through cerebral spinal fluid and can be endoscopically used. The YAG laser is a powerful coagulator of blood which makes the wavelength of great potential for the treatment of vascular lesions.

The greatest concern involving the use of the YAG laser involves its unpredictable depth of penetration and subsequent effect on adjacent vita structures [5, 16]. As opposed to the CO_2 laser where the interaction occurs at the surface which can be easily visualized by the surgeon, the point of interaction from the YAG laser with tissue occurs below the surface.

Although information describing the clinical utility of the YAG laser has been limited, initial information and the experience provided by Tew and colleagues in this volume suggests that this laser will provide the technological means for effectively treating deep seated vascular malformations of the nervous system [5]. Further clinical applications in neurosurgery must await additional investigations involving the use of this laser.

Lasers in Pediatric Neurosurgery

Although the CO_2, Argon, and Nd : YAG lasers have been used in pediatric neurosurgery, the greatest experience has been acquired with the use of the CO_2 laser. The most effective application of this laser in pediatric neurosurgery has been in the removal of neoplasms of the central nervous system [12]. A significant number of childhood brain tumors are located within or adjacent to critical brain structures. Standard surgical techniques may have limitations in such critical areas from excessive manipulation or dissection to achieve adequate access to the tumor.

In addition to intracranial tumors the CO_2 laser has been proven equally effective in treating neoplastic and dysraphic conditions of the spinal cord [4, 6–8]. The following cases will illustrate the clinical utility of the CO_2 laser in pediatric neurosurgery at the Children's Hospital Medical Center, Cincinnati, Ohio.

Clinical Applications

Standard neurosurgical operating techniques were utilized in exposing the pathological condition. A Sharplan 743 laser (LASER INDUSTRIES, LTD., Tel Aviv, Israel), was used in all cases. The laser was attached to an operating microscope with a Contravas stand to provide optimal magnification, illumination, and mobility during the use of the laser.

Case Summaries

Case One—T.N.

An eight-year old white female presented with a two and one-half week history of bilateral temporal headaches, vomiting, and blurred vision. General examination was normal. Neurologic examination revealed bilateral papilledema and nystagmus. Computed tomography (CT) demonstrated a large high attenuation markedly enhancing posterior fossa midline mass with moderate hydrocephalus secondary to obstruction of the fourth ventricle (Fig. 1a).

At surgery the vermis was divided with the CO_2 laser and a tumor which proved to be a medulloblastoma was identified whithin the confines of the fourth ventricle. Using the CO_2 laser slightly defocused at a power setting

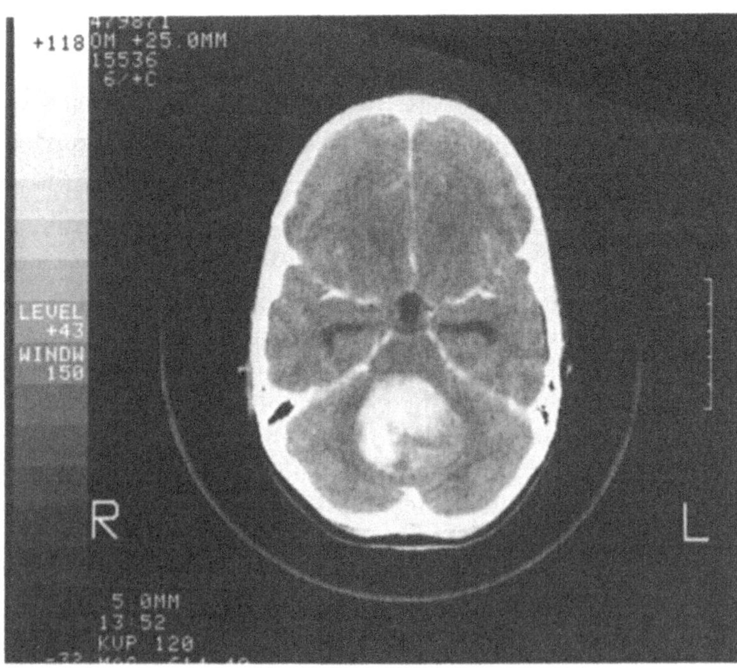

Fig. 1a. Pre-operative contrast enhanced CT of medulloblastoma

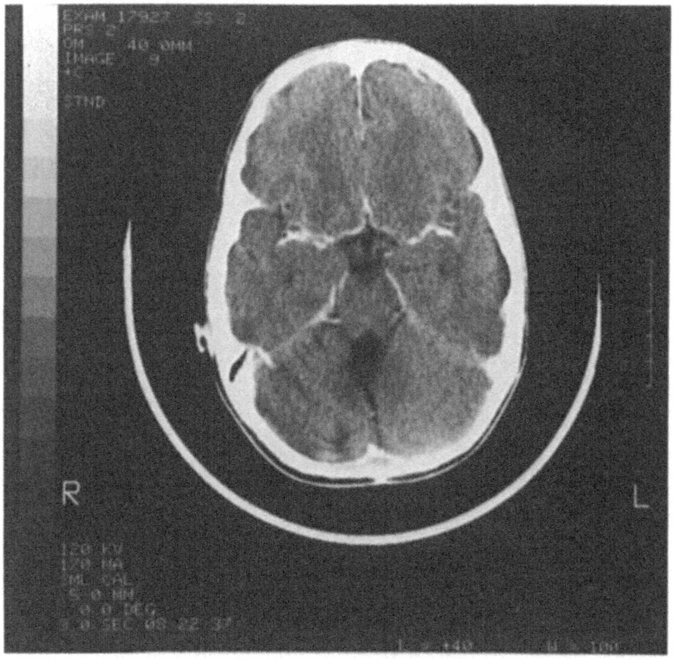

Fig. 1b. Post-operative CT demonstrating complete tumor removal

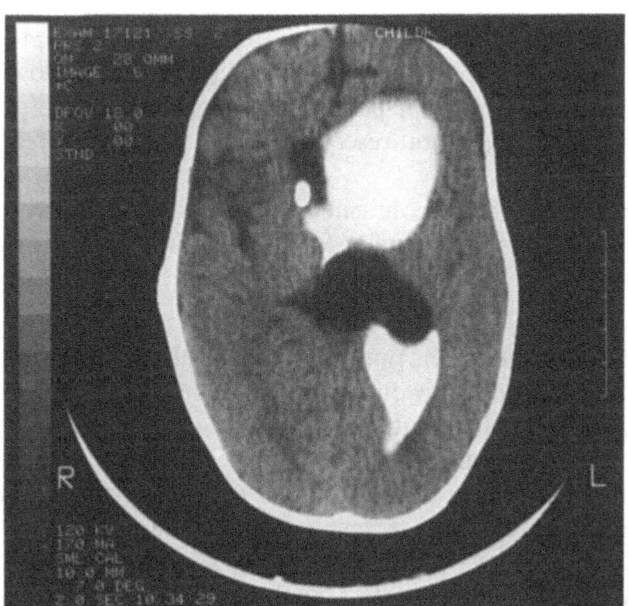

Fig. 2a. Pre-operative CT ventriculogram demonstrating non-communicating intraventricular cyst

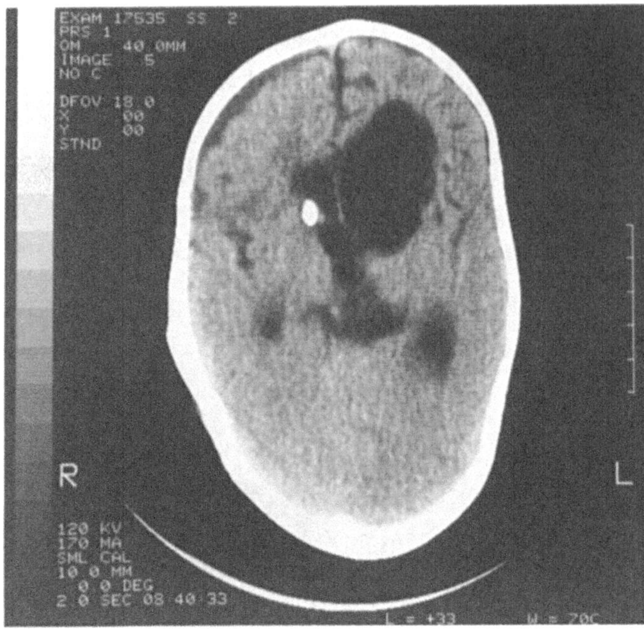

Fig. 2b. Post-operative CT demonstrating diminished ventricular size. Marked clinical improvement accompanied small change in size

of 10 watts, the tumor mass was reduced in size. Wet cottonoid sponges were placed between the cerebellar peduncles and tumor to prevent the spread of heat to the cerebellar peduncles. Progressive dissection with this technique provided gross total resection of the tumor with minimal hemispheric retraction.

Postoperative CT Scan confirmed the gross total resection and the patient was placed upon a course of radiation and chemotherapy (Fig. 1b).

Case Two—E.C.

A three-week old male infant presented with a history of irritability, projectile vomiting, and opisthotonic posturing since birth. Physical examination revealed a bulging fontanelle, poor head control, and prominent scalp veins. Head circumference was at the 98th percentile. Ultrasound examination of the brain revealed dilated ventricles with echogenic fluid. Cerebrospinal fluid obtained from a ventricular tap grew Group B nonhemolytic Streptococcus. Meningitis was successfully treated with antibiotics but hydrocephalus persisted and the child underwent a ventriculoperitoneal shunt. The head circumference remained stable following shunting but a residual left hemiparesis prompted further investigation. Ultrasound examination of the brain suggested loculations of cystic structures

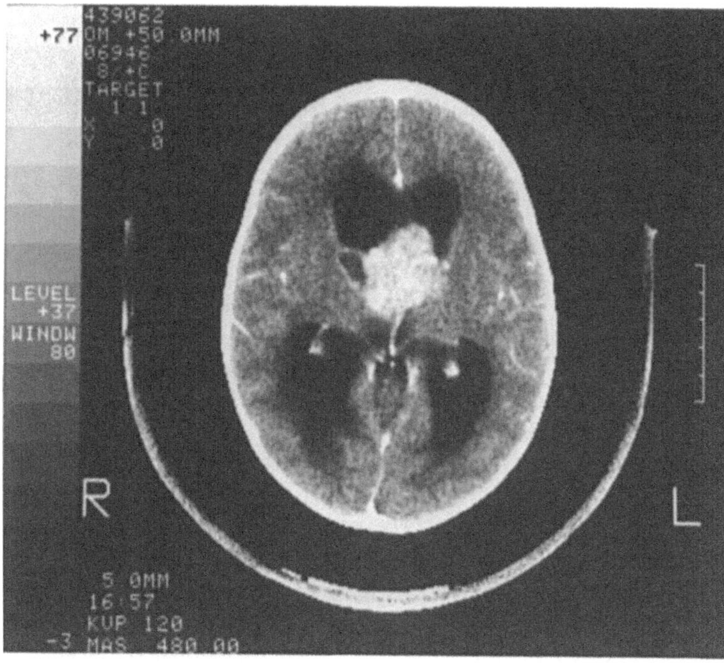

Fig. 3a. Pre-operative CT of intraventricular choroid plexus carcinoma

within the brain. A CT contrasted ventriculogram demonstrated a non-communicating cyst within the body of the left lateral ventricle (Fig. 2a).

The child underwent a transcollosal approach to the ventricular system. Collosal section with the laser provides bloodless dissection and a "no-touch" technique. This maneuver requires the laser beam to be sharply focused and the power setting reduced to a level of 2–5 watts.

Numerous cystic structures within the ventricles were then identified. Using the CO_2 laser at a power level of three watts, the cysts were fenestrated to provide a communication with the ventricular system. One week following the surgical procedure the child was noted to have symmetric movements of the lower extremities and increased movement of the left upper extremity. Decrease in the size of the intraventricular cyst was noted on follow-up CT Scan (Fig. 2b). Cyst fenestration with an endoscopic Argon laser would require less direct operative intervention but perfection of this technique is still lacking [9].

Case Three—A.S.

An 18 month old white female child presented with a one-week history of lethargy, vomiting, and loss of development milestones. The general physical examination was unremarkable. The neurologic examination revealed bilateral sixth nerve palsies and papilledema. A bilobulated mass of high

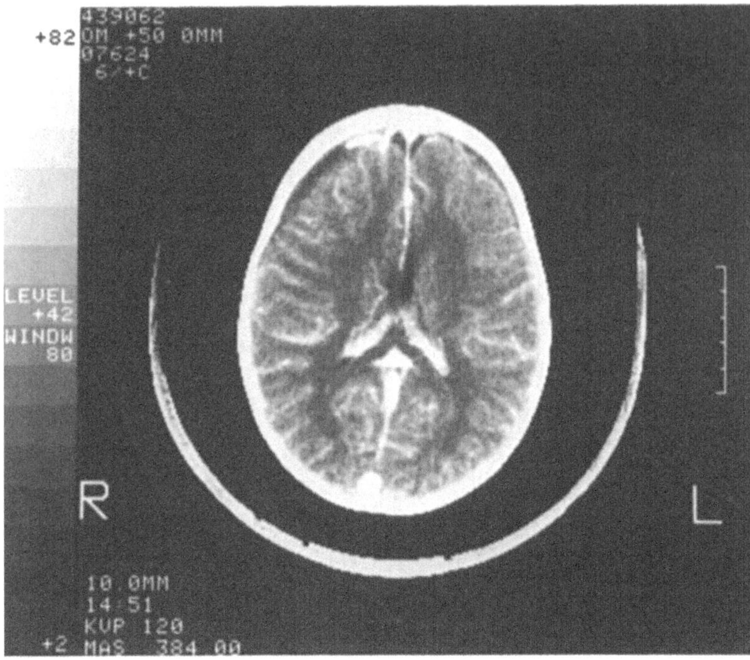

Fig. 3b. Contrast enhanced CT following complete removal of tumor

attenuation was seen on CT (Fig. 3a). The epicenter of the mass was located around the foramen of Monro and extended into the third ventricle. Thickening of the septum pellucidum, extension into the atrium of the lateral ventricles bilaterally was also observed on CT.

A transcollosal approach to the ventricular system was performed. Using the CO_2 laser a two and one-half centimeter incision was made in the corpus callosum. The tumor involved the left lateral ventricle, the septum pellucidum, and infiltrated the corpus callosum posterior to the incision. Using the CO_2 laser the tumor was completely removed including its infiltration of the corpus callosum. Follow-up CT demonstrated normal ventricular size and no evidence of residual tumor (Fig. 3b). Final pathologic diagnosis was a choroid plexus carcinoma. The child underwent radiation and chemotherapy but subsequently expired from disseminated central nervous system metastases.

In this particular case a transcollosal approach provided optimal visualization of this extensive infiltrative neoplasm. The laser permitted complete resection of the tumor with minimal instrumentation through its unique ability to coagulate and vaporize tissue. This feature is valuable when using the transcollosal approach to the ventricular system since the narrow approach limits the amount of direct instrumentation for tumor removal.

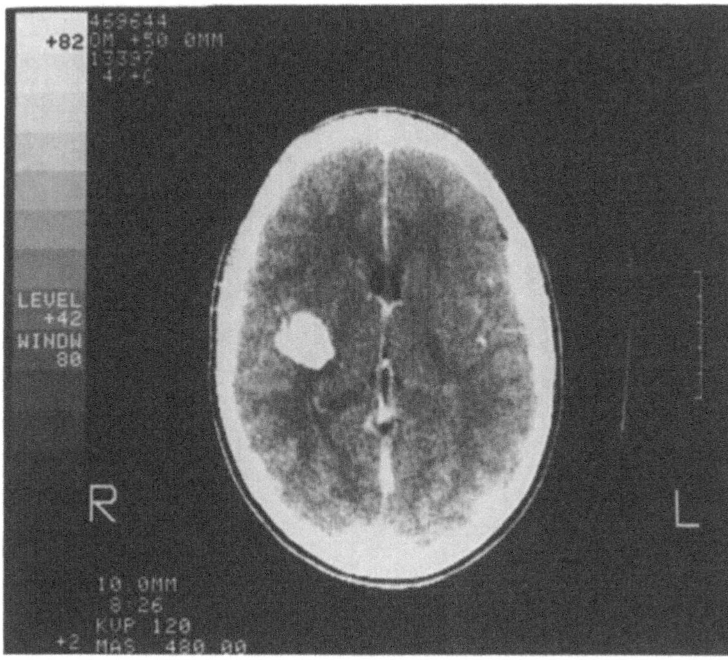

Fig. 4a. Pre-operative CT of calcified Sylvian meningioma

Case Four—T.R.

A 16 year old white male suffered from intractible seizures for several years. The seizures were characterized with tingling and numbness sensation of the left arm and trunk followed by weakness of the left limb and generalized seizures. Various regimens of anticonvulsant medications with therapeutic levels provided inadequate control. CT of the head revealed a well defined calcified lesion buried 4 cm deep in the Sylvian fissure (Fig. 4a).

At surgery the surface of the brain appeared normal but electro-corticography demonstrated sharp wave activity arising from the right post-central region immediately above the Sylvian fissure in the posterior aspect of the superior temporal gyrus. Brain electric stimulation was performed to identify the motor strip. Using the CO_2 laser attached to the operative microscope a linear incision was made along the Sylvian fissure immediately behind the post-central gyrus. A large lobulated solid tumor was located in the Sylvian fissure. Frozen section of the tumor revealed a meningioma. Using the CO_2 laser the tumor was easily vaporized and dissected away from the surrounding brain and branches of the right middle cerebral artery.

Post-resection electrocorticography demonstrated mild residual wave activity in the posterior aspect of the right superior temporal gyrus but did not necessitate further cortical resection.

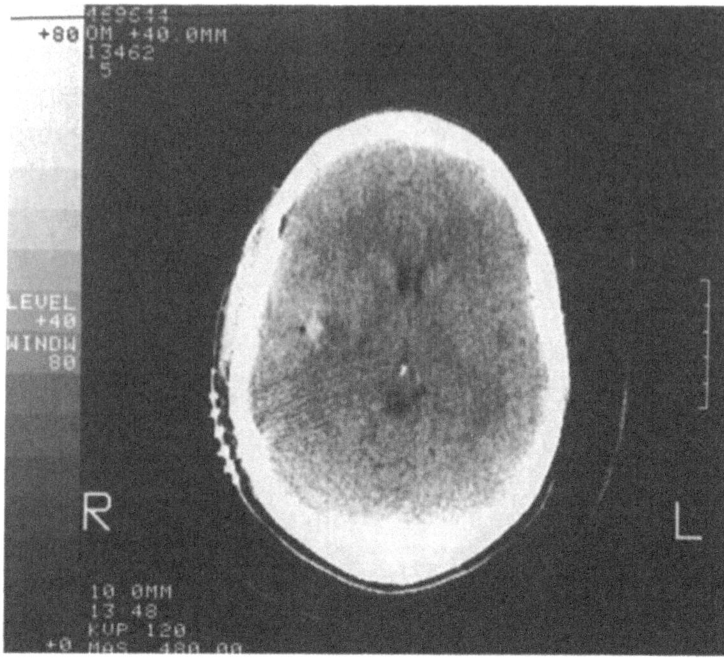

Fig. 4b. Post-operative CT following laser resection

Follow-up CT of the brain revealed complete resection of the tumor (Fig. 4b). Clinical follow-up six months later found the patient to be seizure free on minor anticonvulsant medication.

Case Five—G.S.

This 17 year old white female underwent a previous Harrington rod insertion and bony fusion for severe scoliosis of the spine. Despite Harrington rod insertion and bony fusion the scoliosis progressed and back pain became present while exercising. She was subsequently referred to the Children's Hospital Medical Center for evaluation.

On presentation marked scoliosis of the spine was noted but otherwise the general exam was unremarkable. Neurologic examination revealed intact motory and sensory perception in the lower extremities with intact reflexes and sphincter tones. Only minimal diminution of vibratory sensation was noted in the right lower extremity.

A CT metrizamide myelogram revealed a diffusely swollen cord from the mid-thoracic region to the conus medullaris (Fig. 5). No syrinx was noted on delayed scanning. Magnetic resonance imaging of the spine was suboptimal secondary to the Harrington rod.

At surgery a laminectomy was performed from T-l through the conus

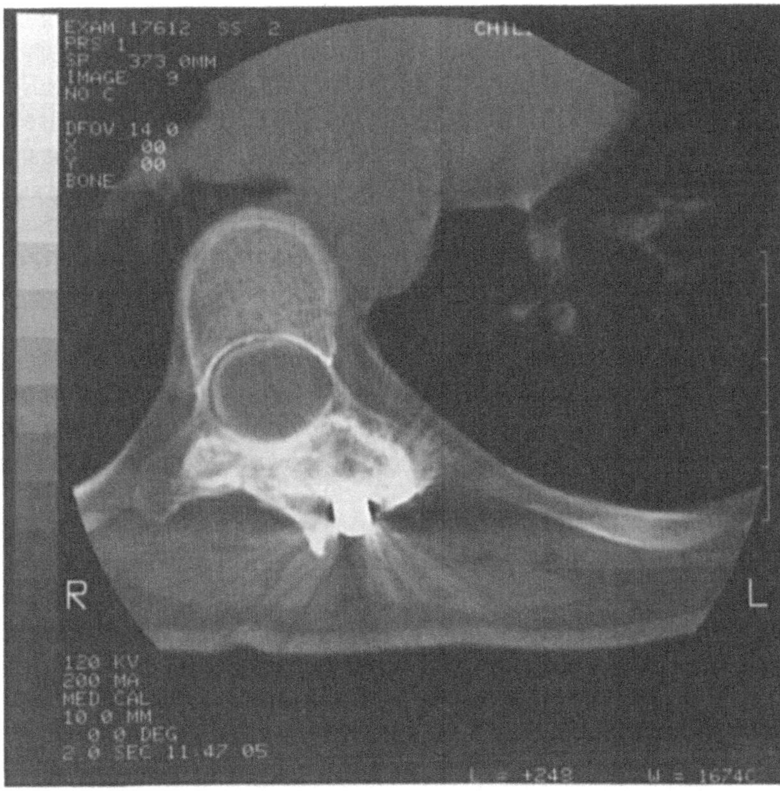

Fig. 5. CT metrizamide myelogram demonstrating enlarged spinal cord

region. A syrinx was identified at T-10 using intraoperative ultrasound. Using the operative microscope with the CO_2 laser a myelotomy was performed at T-10 and carried superiorly for a distance of 14 centimeters. An intramedullary ganglioglioma was resected. Follow-up magnetic resonance imaging revealed complete resection of the tumor.

The laser facilitated a bloodless and precise dissection of a large solid intramedullary tumor. The patient suffered a minimal neurologic deficit which slowly resolved. The laser was effective in this circumstance because of its precision and dissection along the spinal cord-tumor interface without retraction of the spinal cord.

Discussion

As the previous cases have illustrated, the CO_2 laser has been applied to a wide variety of pediatric neurosurgical conditions. The CO_2 laser has been particularly useful in the removal of central nervous system tumors. The ability to vaporize neoplastic tissue without manipulation of adjacent brain or spinal cord structures has permitted safe gross total resection of neoplasms from the ventricles, posterior fossa, and spinal cord. Although large hemispheric tumors may be removed with standard neurosurgical technique, extensive calcifications in such lesions as illustrated in case four are easily vaporized with the CO_2 laser. In addition, the CO_2 laser has been safely applied for the removal of craniopharyngiomas in the suprasellar region and pituitary lesions within the sella turcica without damage to adjacent arteries or nerves. Since laser energy does not interfere with evoked potential monitoring, unnecessary delays are avoided in operative procedures.

The laser has been effective in debulking infiltrative hemispheric and hypothalamic-chiasmatic neoplasms, but the poor demarcation between brain and tumor in such neoplastic conditions has not permitted more extensive resection of such tumors. These observations are similar to those previously reported by Walker [12].

Debulking of tumors is achieved with the laser defocused to a 2–3 millimeter spot size and the power setting adjusted from five to 50 watts. With this technique the tumor folds upon itself and gentle suction will facilitate demarcation of the tumor-brain/spinal cord interface. In critical brain areas wet cottonoid sponges should be applied to nerves, arteries, and adjacent brain tissue to avoid spread of heat to the critical structures or inadvertent transmission of laser energy.

Vessels smaller than one millimeter in diameter are easily coagulated with the CO_2 laser providing hemostasis without accessory instrumentation. More vascularized tumors with larger feeding vessels require use of the YAG laser or bipolar coagulation. The technique for YAG laser removal of arteriovenous malformations and vascular tumors are discussed in other chapters.

For critical brain areas the power of the laser is reduced to a working level of 1–2 watts. With this setting precise dissection with minimal spread of heat provides safe dissection of tumor from the floor of the fourth ventricle, hypothalamus, or tumor adjacent to large vascular channels or nerves in the suprasellar region.

In addition to its treatment of solid neoplasms, the CO_2 laser has been applied to fenestrate isolated cysts within the ventricular system. This technique has facilitated drainage through one shunt tube rather than resorting to complex shunt systems. Powers has recently reported similar success with an Argon laser and endoscope, however, greater experience is needed to evaluate this newer technique [9].

Similar results with the CO_2 laser have been achieved in the treatment of spinal cord conditions. Epstein has reported excellent results at removing intramedullary spinal cord astrocytomas [4]. Neurologic deterioration has not occurred following gross total resection of these tumors. Favorable results have been reported for the treatment of spinal lipomas with or without neurological symptoms. The lipomatous mass which is high in water content can be rapidly vaporized leaving the underlying dysraphic elements intact.

Reports to date including our own observations have been favorable toward the application of the laser to pediatric neurosurgical problems. However, the full extent of this tool for microdissection cannot be fully appreciated until additional experience has been gained with the use of the YAG laser for pediatric vascular malformations and the Argon laser for endoscopic treatment of isolated cysts. Technological advances should further promote even more widespread use of the laser for the treatment of pediatric neurosurgical conditions, a situation where early operative intervention may provide the environment for maximal recovery of neuronal function.

References

1. Boggan JE, Edwards MSB, Davis RL, Bolger CA, Martin N (1982) Comparison of the brain tissue in rats to injury by argon and carbon dioxide lasers. Neurosurgery 11: 609–616
2. Cozzens JW, Cerullo LH (1985) Comparison of the effect of the carbon dioxide laser and bipolar coagulator on the cat brain. Neurosurgery 16: 449–453
3. Edwards MSB, Boggan JE (1984) Argon laser surgery of pediatric neural neoplasms. Child's Brain 11: 171–175
4. Epstein E (1983) Surgical treatment of extensive spinal cord astrocytomas of childhood. Concepts Pediatric Neurosurg 3: 157–169
5. Fasano VA, Urciuoli R, Ponzio RM (1982) Photocoagulation of cerebral arteriovenous malformations and arterial aneurysms with the neodymium: Yttrium-Aluminum garnet or argon laser. Preliminary results in twelve patients. Neurosurgery 11: 754–759
6. James HE, Williams J, Brock W, Kaplan GW, Sang UH (1984) Radical removal of lipomas of the conus and caudina equina with laser microsurgery. Neurosurgery 15: 340–343
7. McClone DG, Hayashida SF, Caldarelli M (1985) Surgical resection of lipomyelomeningoceles in 18 asymptomatic infants. J Pediatr Neurosci 1: 4
8. McLone DG, Nadich TP (1986) Laser resection of fifty spinal lipomas. Neurosurgery 18: 611–615
9. Powers SK (1986) Fenestration of intraventricular cysts using a flexible, steerable endoscope and argon laser. Neurosurgery 18: 637–641
10. Tew JM Jr, Tobler WD (1984) The laser: history, biophysics, and neurosurgical applications. Clin Neurosurg 31: 506–549
11. Walker ML, Storrs BB, Goodman SG (1983) Use of the CO_2 laser for surgical excision of primary brain tumors in children. Concepts in Pediatric Neurosurgery 3: 297–315

12. Walker ML, Storrs BB (1985–86) Lasers in pediatric neurosurgery. Pediatr Neurosci 12: 23–30

Address for correspondence: Kerry R. Crone, M.D., Assistant Professor, Children's Hospital Medical Center, Department of Pediatric Neurosurgery, Pavilion Building, Room 2–9, Elland and Bethesda Avenues, Cincinnati, OH 45229-2899, U.S.A.

The Use of Contact Laser in Neurosurgery. Clinical and Experimental Data

V. A. Fasano and **R. M. Ponzio**

Institute of Neurosurgery, University of Turin, Italy

Introduction

Recent technological improvements have resulted in the introduction of contact probes which have improved laser fiberoptic delivery system and substances able to transmit laser light. The sapphire tips introduced by Daikuzono and Joffe in 1985 [1] are currently used to deliver more than 90% of laser light, causing no tissue adhesion phenomena. This is an artificial crystal which is hard enough and has enough mechanical resistance to prevent breaking of the tip. Lateral irradiation, which results in 30–40% of the total power delivered in non-contact probes, is avoided, thus providing substantial reduction in laser energy requirements. The contact tips are so efficient that very low thermal energy is required for a sharp incision, because of such rapid tissue vaporization. The greater divergence of the beam limits the thermal effect, thereby minimizing the lesion. The contact probes are suitable for both freehand and endoscopic application.

Instrumentation

Contact laser systems can be adapted to non-contact laser systems. However, it must be pointed out that stability in delivery of low power energy is required to avoid melting of the sapphire tip. The sapphire connection with the optic fiber is easily carried out simply confronting the two surfaces. A cooling system avoids tip overheating and assures the removal of smoke and gases produced by tissue burning.

Two sets of probes are available; one for cutting and one for hemostasis. Diameter of the distal end of the tip accounts for the diameter of the beam at the target and as a general rule the thinner probes require lower energy settings to achieve corresponding tissue effects. Hemostatic efficiency is higher with the larger sapphires. The incisional capability depends on the laser power input and the diameter of the probes. The following probes are currently used. 1) Sapphires with a 0.2 up to 1.2 mm diameter and with 9 up to 19 mm length. The larger tips are used for dissection in cavities; 2) A sapphire with a 0.05 mm diameter which is employed in microsurgery. There is a frosty surface sapphire which delivers energy from the tip and lateral surface that is also used in highly vascularized tissues.

Laser sources used in contact delivery systems are Nd : YAG and argon.

Neodymium : YAG is a solid state laser, made from a crystal yttrium aluminum Garnet with incorporated (ion doped) Nd 3+ of certain concentrations; these ions are excited by the absorption of light energy. The beam is emitted on a near infrared range with a wavelength of 1.06 microns. Guide beam is a coaxial 2 w He-Neon laser.

Argon is an ion-gas laser. Excitation is obtained by producing an electrical discharge at a very high current density in ionized gas. The beam is visible, ranging in the blue-green part of the spectrum with a wavelength of 488–514 nm. More than 70% of the energy is selectively absorbed by hemoglobin. Transmission is through fiberoptics. When used at non-contact, backwards and forwards scattering are evident in both lasers, incision being wide and faintly wedgeshaped. Hemostasis is sufficient for veins and arteries up to 1 mm.

The KTP/532 laser is a pure green laser wavelength of 532 nm, produced by a potassium titanyl phosphate crystal. Theoretically due to the higher absorption of red oxygenated hemoglobin by the green light (about 95%) hemostasis is enhanced. However, penetration depth is influenced by chromophore density, i.e., tissue vascularity. This wavelength cuts, coagulates and vaporizes more efficiently, with less carbonization and smoke than with CO_2. The surgical system is a frequency-doubled Nd : YAG laser by

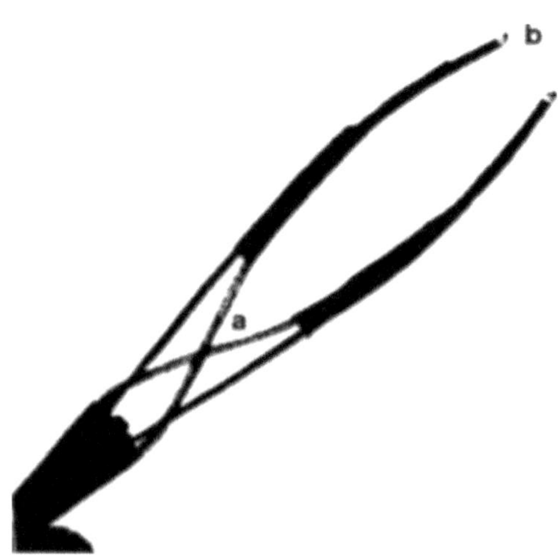

Fig. 1. Bipolar forceps. Prototype (*a*) optic fiber, (*b*) sapphires

a KTP crystal to produce the 532 nm output. An optical fiber cable stored in the console transmits the laser light to a remote optic coupler. This coupler enables sterilized delivery devices to be attached close to the operating field allowing direct tissue contact without fiber modification.

Contact lasers operate like a usual scalpel and can be combined with operating microscope at the 0.05 mm tip with no visual encumbrance. To prevent melting of the point the end of the sapphires must be kept in contact with the tissue during incision. Tissue adhesion must be carefully avoided since carbonization of tissue debris produces direct thermal effects. Research is in progress to develop a laser forceps. The instrument consists of two sapphires at the end of the branches linked to two optic fibers. The laser energy is divided into two branches by a splitter and a low-power Nd:YAG irradiation is delivered (1–2 watts) (Fig. 1).

Experimental Data

Experimental studies on contact Nd:YAG laser include the possibility of a direct thermal effect. As a matter of fact, when tissue adhesion is avoided, thermal delivery averages 1/1,000 of the total power (Fig. 2). Thermal effect on surrounding tissues is limited. The table shows the extension in mm of thermal diffusion, caclulated for 42 °C temperatures (threshold of cellular damage), 60 °C (thermal coagulation), 100 °C (permanent cellular damage), is drastically reduced as far as conventional laser systems are concerned, and the equivalent emission powers from a surgical point of view are considered (Fig. 3).

P-laser (Watt)	P-thermal (milliwatt)	Temperature (K)
23.5	21	1,800
16.5	15	1,650
11	9	1,450
8.5	7	1,350

Note: P-thermal is normalized to the same angular aperture of the outcoming laser radiation. Irradiation time: 10 seconds

Fig. 2. Dependence of the Tip Temperature (T) and corresponding thermal radiated power (P-thermal) on the total radiated laser power (P-laser). The figure reports the amount of the power responsible for a direct thermal effect, considering various values of total transmitted power. For a total power of 23.5 watts, the thermal effect corresponds to a power of only 21 milliwatt (less than 1/1,000 of the power delivered), for a total power of 16.5 watts, to a power of 15 milliwatts and so on. On the left are reported for each value of power the temperature of the beam at the focus

P \ T	42 °C	60 °C	100 °C
25	14.6 (9.9)	10.0 (5.3)	5.9 (1.2)
28	15.2 (10.4)	10.6 (5.8)	6.4 (1.7)
31	15.6 (10.9)	11.0 (6.3)	6.9 (2.2)
15	11.0 (6.3)	6.4 (1.7)	2.3 (—)
18	11.8 (7.1)	7.3 (2.5)	3.2 (—)
21	12.6 (7.9)	8.0 (3.3)	3.9 (—)

Fig. 3. Thermal effect in the cerebral tissue after contact and non-contact irradiation at equivalent emission powers in the hypothesis of missing vaporization and loss of heat by vaporization (values in brackets). With contact laser irradiation at 18 watts, tissular temperatures of 100 °C are limited to the area of laser impact, temperatures of 60 °C and 42 °C have been recorded at 2.5 mm and 7.1 mm from the target.
With non-contact laser, however, tissular temperatures of 100 °C–60 °C and 40 °C are recorded respectively at 1.7 mm–5.8 mm and 10.4 mm from the target

Furthermore, it has been shown that as far as contact lasers are concerned (0.5 mm sapphire's tip diameter), 80% of delivered power is emitted in a 90° cone shaped beam and that the power density at the tip corresponds to 2,500 watt/cm^2 for a delivered laser power of 6.5 watt (Fig. 4). In the conventional non-contact laser, 80% of energy is focused in a 12° cone and the power density at the end of the fiber is about 400 watt/cm^2 for a delivered power of 13 watt (Fig. 5). This accounts for the vast energy at the target and instantaneous tissue evaporation produced.

Owing to the greater divergence of the beam, the power density reduces itself to 40 watt/cm^2 at a distance of 2 mm from the tip of a sapphire, while in non-contact laser at a distance of several cm from the end of the fiber, power density still corresponds to 400 watt/cm^2. This is responsible for less tissue damage in depth after contact irradiation.

Histological studies made on different tissues after incision with contact Nd:YAG show the absence of carbonization, whereas the area of coagulative necrosis spreads for 15–40 micron only into extracranial tissue (temporal muscle) and for 100–200 micron into cerebral tissue [2].

The ultrastructural study of the effects on adjacent tissues (tissue lesions 1.5–3 mm away from the center of the lesion) shows different pictures after Nd:YAG and Argon irradiation.

The Argon causes slight modifications in erythrocytes, endothelial cells and neurons in the form of vacuolization along with a total preservation of vessel perviety and tissue structure (Fig. 6a, b).

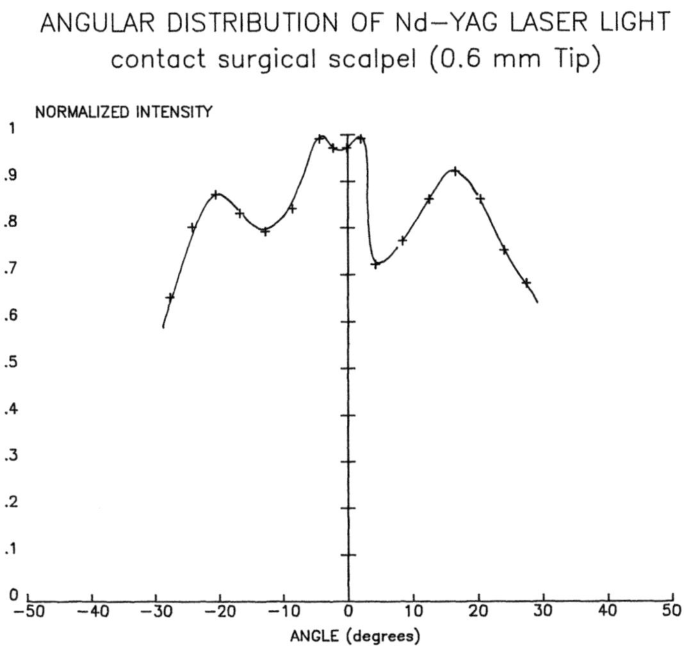

Fig. 4

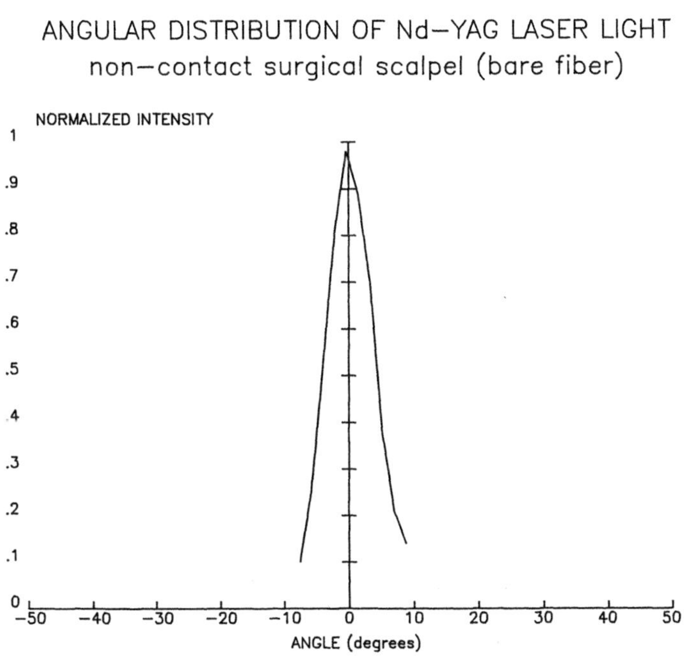

Fig. 5

Figs. 4 and 5. Morphology of the emission beam in contact (4) and non-contact (5) irradiation

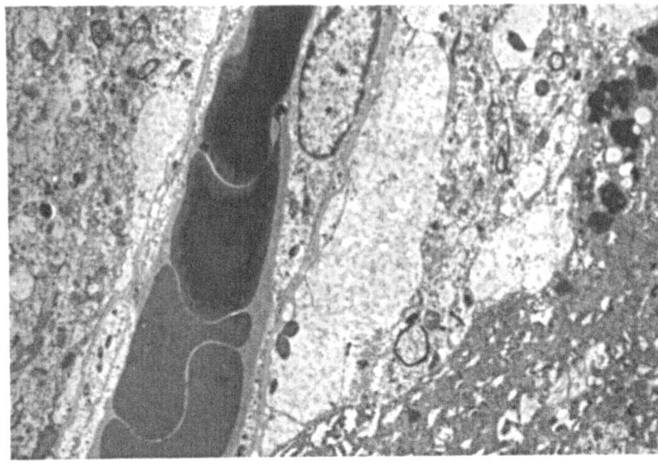

a

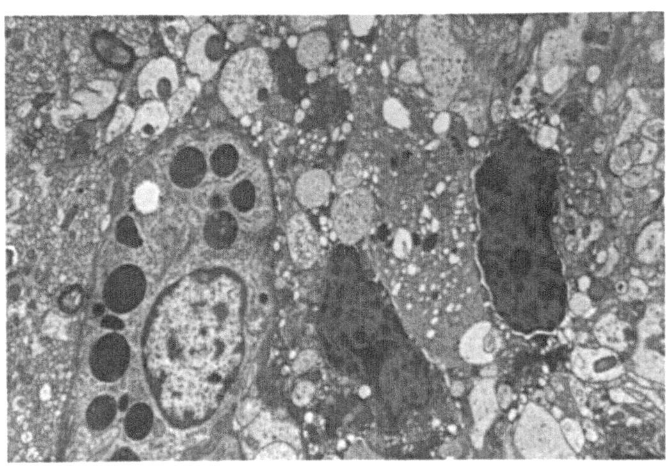

b

Fig. 6a, b. Argon laser: This source produces minimal tissual changes, consisting in appearance of vesicles into endothelial cells and basement membrane. a) Shows a cerebral capillary with the endothelial cells, basement membrane and erythrocytes into the lumen. Only vesicles into basement membrane are evident. b) Shows normal cerebral tissue; nuclei, granules of pigment, astrocyte pedicles and myelinic sheaths are evident. The laser-induced damage is limited to the appearance of several intracytoplasmic vesicles

Nd:YAG laser, instead, causes clear damage in the microcirculation and tissue with nuclear and cytoplasmic lesions being present in the astrocytic pedicles and endothelial cells (Fig. 7a, b) while in some samples erythrocytes are aggregated.

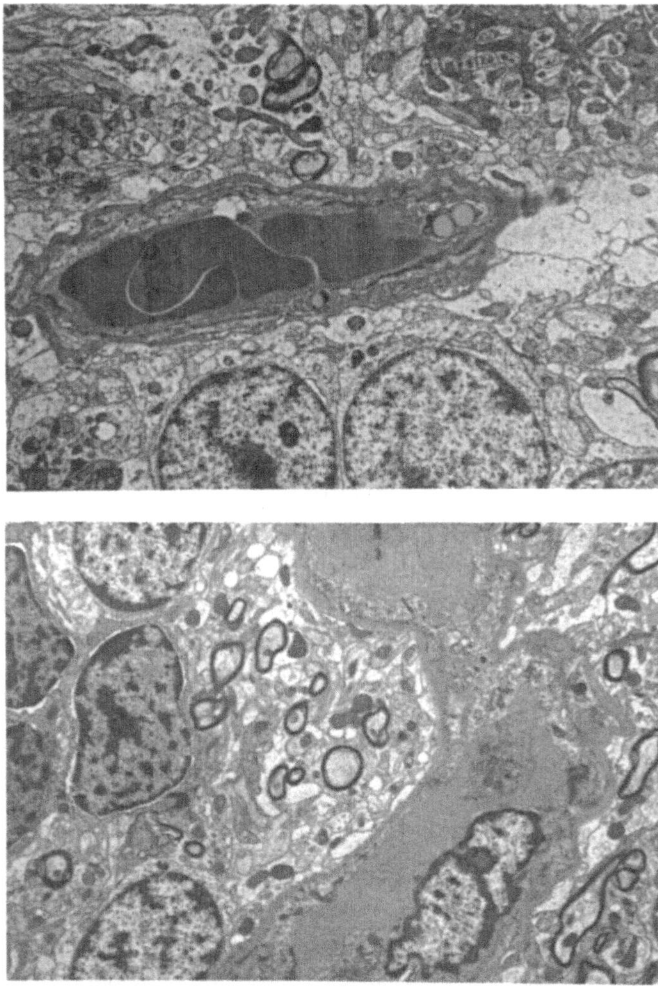

a

b

Fig. 7a, b. Nd: YAG laser: This source produces marked histological changes, consisting of swelling of cells and basement membrane and in appearance of intracytoplasmic vesicles. a) Shows a cerebral capillary. The structure is preserved. Swelling of the endothelial cells and vesicles into the basement membrane are observed. b) Shows the cerebral tissue. The tissue architecture with nuclei, granules of pigments and myelinic sheaths are observed. Astrocyte pedicles are swelled and several intracytoplasmic vesicles are detectable

Surgical Technique

Dissection of the lesion from the surrounding healthy tissue is made with a 0.05–0.6 mm sapphire tip under the operating microscope.

Cutting the cortex is performed by simply grazing the tissue, a maneuver which can be progressively repeated. The width of the incision can be regulated between 0.2 to 1.2 mm. Cutting of the white matter is performed with the tips of large diameter at 10 to 15 watts of power. Different tips and powers are employed on tumor surfaces, depending on the firmness of the tumor. The sapphire tips make incisions between 0.8–1.2 mm in hard tissues, with a power ranging for 12 to 18 watts. Incision of fibrous structures is performed with 0.2 to 0.6 sapphire tips at power less than 10 watts. Maneuverability is easy and control of veins complete.

Coagulation of smaller vessels can be easily performed using a sapphire of 0.1–1 mm whose tip and lateral portion are frosty. Energy is delivered at low power (1–4 watts). Vessels up to 1 mm can be segmentally occluded by gently pressing the vessel in order to produce a gluing of the walls. This maneuver can be performed by the flat tip or using the larger tip sapphires usually employed for cutting.

Results

Our experience and results of 154 cases of intra-axial and extra-axial tumors are listed in the tables (Figs. 8 and 9).

The main indications for use of contact lasers are the following:

a. Extra-axial tumors: basal meningiomas, falx and parasagittal meningiomas.

Main procedures are:

1) peeling of the tumor from the surrounding neurovascular structures (0.05 mm sapphire with the aid of the operating microscope). Occurrence of side effects is avoided because surgical procedure is limited to the target, all energy being focused on the tip.

2) Cutting of the implant of the tumor from fibrous structures (falx, tentorium, dura matter). The last indication is the most gratifying, making the contact laser the instrument of choice.

b. Intra-axial tumors: hemispheric cortico-subcortical gliomas and midline gliomas.

Main procedures are:

1) Gyral (cortical gliomas) or intergyral incision after opening of the arachnoid; the approach is easier since use of additional instruments is avoided and encumbrance of the operating field reduced.

2) Dissection of the tumor from the ventricular walls; this is easily carried out even in the presence of fluids.

3) Hemostasis of the smallest vessels of the white matter.

Areas where the use of the contact laser is less effective.

1) Hemostasis is incomplete in bleeding vessels and in arteries larger than 1 mm. In these cases bipolar forceps are used. Laser forceps are predicted to substitute bipolar forceps in order to improve maneuvers on vessels since the graduation in energy delivery and the selective effects on walls and lumen such as sticking and charring are avoided.

Series of cases: 154 Pathology	N° of cases
Cortico-subcortical gliomas	45
Deep-seated gliomas	15
Convexity endotheliomas	15
Parasagittal meningiomas	14
Falx and tentorial meningiomas	10
Sphenoid wing meningiomas	6
Posterior fossa meningiomas	5
Ependimomas	5
Cerebellar hemangioblastomas	1
Cerebellar Spongioblastomas	15
Metastases	13
Abscesses	7
Radionecrosis	3

Fig. 8

Maneuvers performable with contact laser	Advantages
Incision of cortex and white matter	Reduced width of cut (0.05 mm) Complete hemostasis
Peeling of residual tumors from the adjacent tissues and dissociation of nervous-vascular structures	Reduced width of cut (0.05 mm), reduced lateral thermal diffusion, reduced side effects and mechanical trauma
Contact hemostasis of arteries up to 1 mm and veins up to 3 mm	Low power required for vessel closure. Use of bipolar forceps limited to bleeding vessels and larger arteries
Resection of tumor implant	Rapid and bloodless resection of dural and fibrous structures (falx and tentorium)

Fig. 9

Figs. 8 and 9. Series of cases and advantages of the contact probes in the surgical procedure

2) Cutting of hard tissues (thicker meningiomatous capsulae, calcified tissues) is more difficult requiring larger sapphires and higher power (up to 15 watts).

3) High occurrence of tip breaking when irradiation is kept outside the tissue, even when the stability of energy delivered is assured. Research is in progress to find stronger substances for contact probes.

4) Requirement of constant visual control of the sectioned planes in order to avoid tissue adhesion to the tip.

Conclusions

As compared with the traditional instruments, the laser scalpel is the best instrument of dissection and cutting of nervous tissue and fibrous structures, showing definite advantages of the electric knife and bipolar forceps. The procedure is more precise and gradable, regularly making an incision of 0.05 mm in width possible. With bipolar forceps the section is more traumatic, irregular and wider whereas the hemostasis of the smaller vessels of the white matter often causes a remarkable extension of tissue damage.

Contact lasers offer some remarkable advantages even as compared to non-contact systems.

1) The coagulating properties of the Nd : YAG are combined with cutting capabilities previously offered by CO_2 laser only.

2) The tactile feed-back the operator has lost with the use of non-contact lasers is restored.

3) The efficiency of the system is greater. Due to the geometric and optic characteristics of the sapphire, all the energy is focused on the tip and the divergence of the beam at the exit is high. This accounts for the lower power settings required and for reduction of sidewards irradiation. Size of laser apparatuses are consequently reduced.

4) The contact tip allows powers of 25 watts or less to cut, coagulate, vaporize and deliver interstitial irradiation because of improved focusing and lesion predictability.

5) Smoke and gas production is limited.

6) The hemostasis by segmentary coagulation is effective for vessels up to 1 mm. Maneuvering is easier even in pulsatile arteries and blood loss is reduced.

7) The width of incision is graduable, thinner and more precise than the CO_2 laser.

8) The dissection is performed with higher selectivity and minimal thermal diffusion. The handiness of the tool becomes particularly important in the separation of tumor lesions from the surrounding tissue surfaces when they are irregular.

9) Greater safety in cutting or dissecting cavities because of lesion predictability.

In laser surgery a proper selection of contact and non-contact irradiation is now required. Contact lasers are indicated for dissection and cutting in depth while non-contact lasers are most useful for vaporization and removal of small thin lesions on the surface. The two techniques can be combined quite satisfactorily.

Summary

The rationale behind the employment of the laser in surgery is its ability to focus high energy levels on restricted areas with minimal side-scattering and diffusion of heat. Dissociation of tumors from adjacent structures after debulking by ultrasonic aspirator is performed with contact laser systems (Argon or Nd : YAG). Lateral irradiation is greatly reduced with this technique, all the power being delivered at the distal end. Moreover, when Nd : YAG is used, the source's coagulating powers are preserved. Spectrophotometric studies have shown that direct heat makes virtually no contribution to the production of the lesion since it corresponds to less than one-thousandth of the total energy delivered. Contact probes are suitable for hemostasis, either in surface or in cavities (flat-tip); cutting of poorly vascularized tissues using a 0.05 mm tip designed for microsurgery or a frosty tip with emission of the beam both at the tip exit and along the lateral aspect of the crystal to improve hemostasis. At the beginning of the procedure (incision of cortex) we use low power settings to enhance the hemostatic capabilities. The cut of white matter (poorly vascularized) is then performed using higher power. The use of bipolar forceps is thus limited to the coagulation of bleeding vessels. The vaporization of tumor renmants is performed without contact. The two techniques are thus complementary.

References

1. Daikuzono N, Joffe SN (1985) Sapphire probe for contact photocoagulation and tissue vaporization with the Nd : YAG laser. Medical Instrumentation 4: 173–178
2. Diaz FG, Dujovny M, King PK, Chason J, Ausman JI, Malik G, Berman SK (1986) Use of the contact Nd : YAG laser scalpel in neurosurgery. Proceedings of the 4th general and scientific meeting of the LANSI (Laser Association of Neurological Surgeons International), Venice, in press
3. Fasano VA, Peirone SM, Ponzio RM, Lanotte MM, Merighi A (1986) Effect at the periphery of the laser lesion in human brain and its tumors after CO_2, Nd : YAG and CO_2 high peak pulsed radiation. Lasers Surg Med 3: 308–317
4. Karnowsky MJ (1965): A formaldehyde-glutaraldehyde fixative of high osmolarity for use in electron microscopy. J Cell Biol 27: 137A–138A
5. Jain KK (1985) Complications of the use of the Nd : YAG laser in neurosurgery. Neurosurgery 16: 759–762

Address for correspondence: Prof. V. A. Fasano, M. D., Ph. D., Institute of Neurosurgery, University of Turin, Via Cherasco 15, I-10126 Torino, Italy.

Subject Index

GPSR Compliance

The European Union's (EU) General Product Safety Regulation (GPSR) is a set of rules that requires consumer products to be safe and our obligations to ensure this.

If you have any concerns about our products, you can contact us on ProductSafety@springernature.com

In case Publisher is established outside the EU, the EU authorized representative is:

Springer Nature Customer Service Center GmbH
Europaplatz 3
69115 Heidelberg, Germany

Batch number: 10584556

Printed by Printforce, the Netherlands